REVISE AQA GCSE (9–1)

Biology

REVISION WORKBOOK

Higher

Series Consultant: Harry Smith

Author: Nigel Saunders

Contents

AQA publishes Sample Assessment Material and the Specification on its website. This is the official content and this book should be used in conjunction with it. The questions have been written to help you practise every topic in the book. Remember: the real exam questions may not look like this.

Microscopes and magnification

1 A student uses a light microscope. The eyepiece lens has a magnification of ×10 and the objective lens has a magnification of ×5. What is the total magnification? Tick **one** box.

Always answer multiple-choice questions, even if you don't actually know the answer or can't work it out.

×2 ☐ ×5 ☐ ×15 ☐ ×50 ☐ **(1 mark)**

2 Scientists use light microscopes and electron microscopes to study cells. Describe how these two types of microscope differ in their magnification and resolution.

Guided

The magnification of a light microscope is usually ....

than the magnification of an electron microscope. The level of detail seen with a

light microscope is .... than that with an electron

microscope because its resolution is .... **(2 marks)**

3 The photo is an electron micrograph of part of a human liver cell.

Maths skills An answer to 1 significant figure is sufficient.

mitochondrion

nucleus

2 µm

(a) Estimate the length of the mitochondrion.

....

.... **(1 mark)**

(b) Estimate the diameter of the nucleus.

.... **(1 mark)**

4 A bacterial cell is 3 µm long. Its image in a microscope is 1.5 mm long.

(a) Calculate the magnification of the microscope when it forms this image.

Maths skills 1 mm = 1000 µm

....

.... **(2 marks)**

(b) The magnification is adjusted to ×750. Calculate the new size of the image of the bacterial cell.

....

....

.... **(1 mark)**

5 Explain how electron microscopy has increased scientists' understanding of sub-cellular structures.

....

.... **(2 marks)**

Animal and plant cells

1 Which of the following structures is the site of protein synthesis in a cell? Tick **one** box.

cell membrane ☐ mitochondrion ☐ ribosome ☐ nucleus ☐ **(1 mark)**

2 The diagram shows a type of cell.

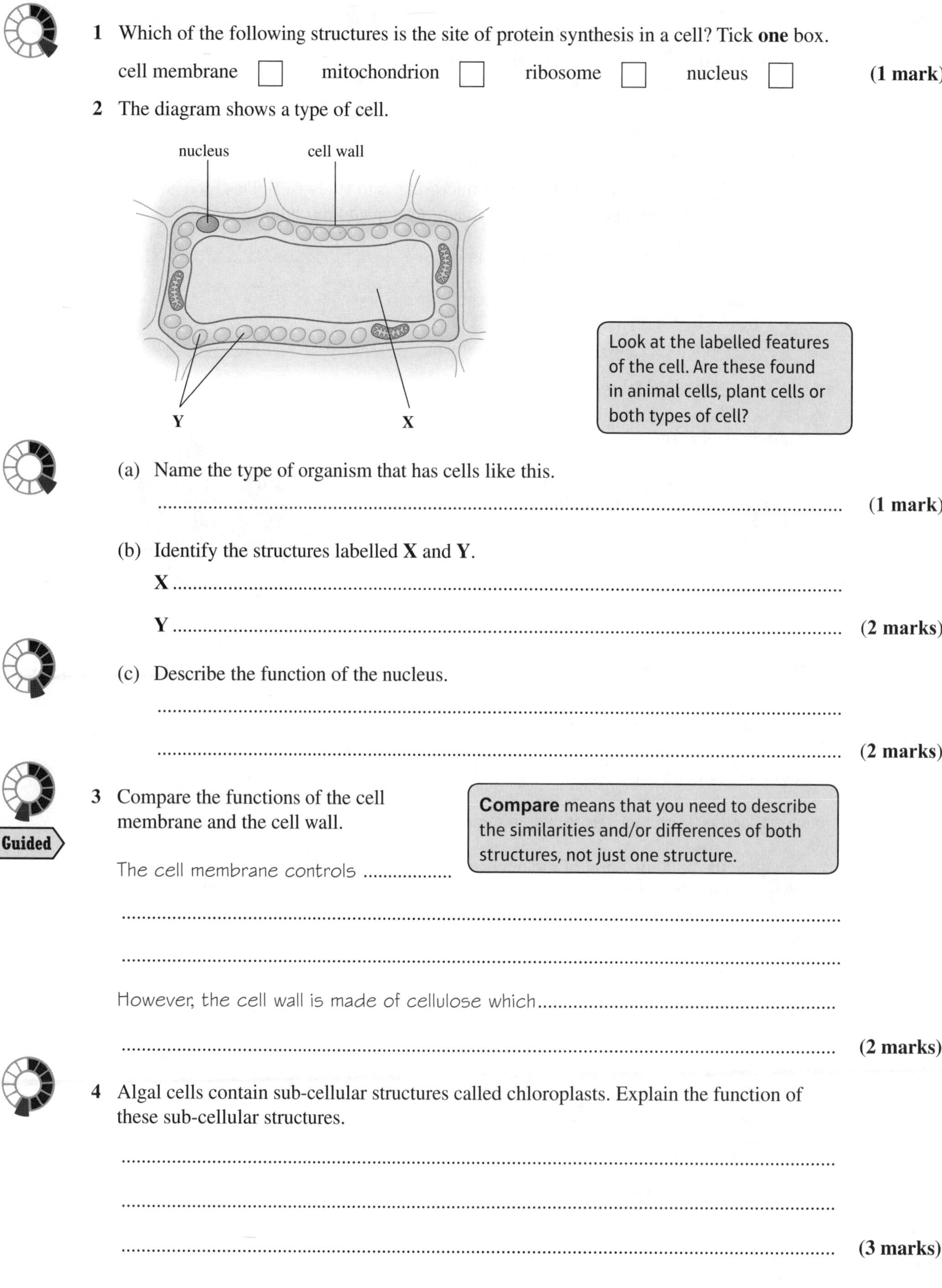

Look at the labelled features of the cell. Are these found in animal cells, plant cells or both types of cell?

(a) Name the type of organism that has cells like this.

.... **(1 mark)**

(b) Identify the structures labelled **X** and **Y**.

X ....

Y .... **(2 marks)**

(c) Describe the function of the nucleus.

....

.... **(2 marks)**

Guided

3 Compare the functions of the cell membrane and the cell wall.

Compare means that you need to describe the similarities and/or differences of both structures, not just one structure.

The cell membrane controls ....

....

....

However, the cell wall is made of cellulose which ....

.... **(2 marks)**

4 Algal cells contain sub-cellular structures called chloroplasts. Explain the function of these sub-cellular structures.

....

....

.... **(3 marks)**

Eukaryotes and prokaryotes

1 Animal cells are examples of eukaryotic cells. Bacterial cells are examples of prokaryotic cells. Complete the table to show which features are present in these cells. Place a tick (✓) in each correct box to show where a feature is present.

	Animal cells	Bacterial cells
Cytoplasm		
Cell membrane		
Cell wall		
Nucleus		

(4 marks)

2 Describe how the genetic material is arranged in prokaryotic cells, such as bacterial cells.

Guided

The chromosomal DNA is arranged to form a ....

Some bacterial cells also contain .... **(2 marks)**

3 Write the following measurements in order of increasing size.

1 cm	50 μm	100 mm	200 nm

(Smallest) .... (largest) **(1 mark)**

4 Convert the following measurements to metres in standard form.

Maths skills Numbers in standard form are written as: $A \times 10^n$

- A is a number greater than or equal to 1, and less than 10
- n is a power of 10

(a) 0.0022 m

.... **(1 mark)**

(b) 0.45 mm

.... **(1 mark)**

(c) 97 μm

.... **(1 mark)**

5 The diameter of a liver cell is 2.5×10^{-5} m. The diameter of a bacterial cell is 2.0×10^{-7} m.

(a) Calculate how many times larger the liver cell is than the bacterial cell.

....

....

.... **(2 marks)**

(b) Give the order of magnitude of the diameter of the bacterial cell.

The order of magnitude of the diameter of the liver cell is –5.

.... **(1 mark)**

(c) Determine how many orders of magnitude larger the liver cell is than the bacterial cell.

.... **(1 mark)**

Had a go ☐ Nearly there ☐ Nailed it! ☐

Specialised animal cells

1 The diagram shows a sperm cell.

acrosome
nucleus
mitochondrion
tail

Draw **one** line to link each structure to its correct function.

Structure	Function
acrosome	releases energy for the cell
nucleus	allows cell to move
mitochondrion	carries genetic information
tail	releases enzymes to aid entry to an egg cell

(4 marks)

2 The diagram shows a human red blood cell.

Red blood cells contain haemoglobin. This protein binds to oxygen so it can be carried in the bloodstream.

Suggest a reason why the red blood cell does not contain a nucleus.

.... **(1 mark)**

3 The diagram shows a nerve cell.

Nerve cells connect with other cells, and carry electrical impulses between distant parts of the body.

Explain how the nerve cell is adapted to its function.

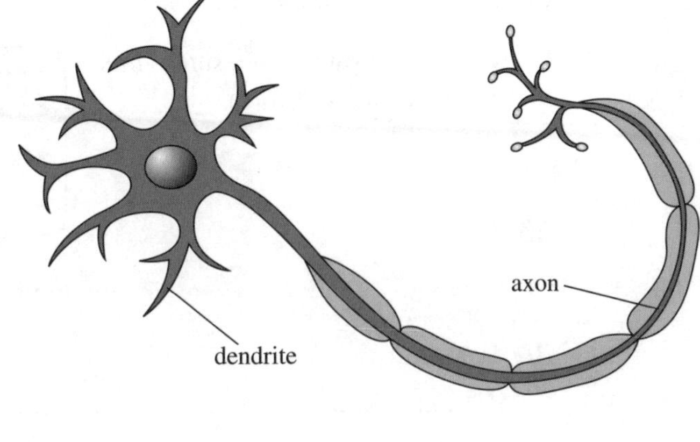

Explain means that you need to describe each labelled part and then say how it allows the cell to do its job.

....

....

....

.... **(4 marks)**

4 Describe what happens as cells differentiate in animals.

Most types of animal cells differentiate at an .... stage.

As a cell differentiates, it acquires different ....

that allow it to ....

Cell division in mature animals is mainly restricted to .... **(4 marks)**

Specialised plant cells

1 The diagram shows a longitudinal section through phloem tissue. It consists of sieve cells and companion cells. Phloem is a transport tissue that carries dissolved sugars through a plant.

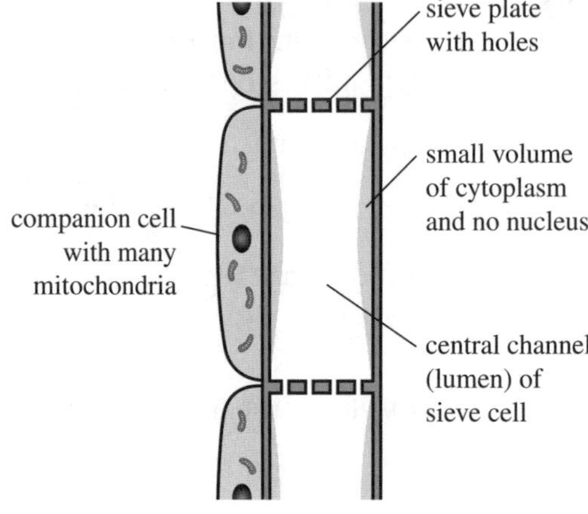

Choose from the labelled features on the diagram in your answers.

Identify the feature which:

(a) provides a lot of room for a central channel

.... **(1 mark)**

(b) allows liquids to flow from one cell to the next

.... **(1 mark)**

(c) transfers energy for active transport

.... **(1 mark)**

2 Xylem tissue is a transport tissue that carries water and mineral ions from the roots to the rest of the plant. The diagram shows part of a xylem vessel.

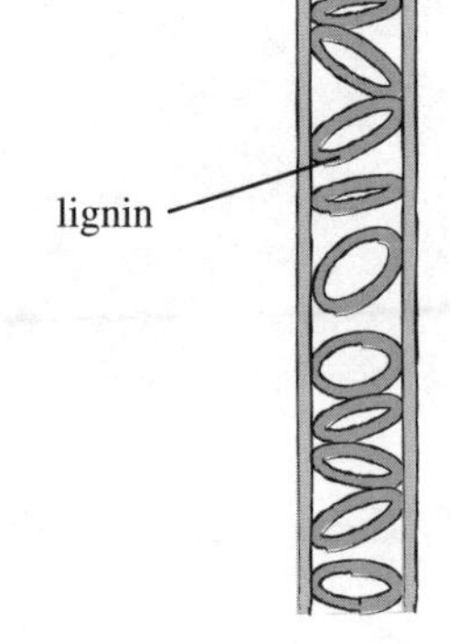

Guided

(a) Suggest a reason why there are no end walls between individual xylem cells.

Xylem tissue consists of hollow tubes formed by dead xylem cells. There are no end walls so that

....

.... **(2 marks)**

(b) Lignin is a tough substance that builds up in xylem cells.

(i) Give a reason why the presence of lignin is important to the function of xylem tissue.

.... **(1 mark)**

(ii) Lignin is insoluble in water. Explain why this is important for the function of lignin in xylem tissue.

....

.... **(2 marks)**

3 Describe how, during the life of the organism, the ability of plant cells to differentiate differs from the ability of animal cells to differentiate.

What is different about when the cells can differentiate?

....

.... **(2 marks)**

Had a go ☐ Nearly there ☐ Nailed it! ☐

Using a light microscope

Some designs of microscope use a mirror instead of a lamp.

1 Describe the function of the following parts of a light microscope.

(a) the lamp

.... **(1 mark)**

(b) the stage with clips

.... **(1 mark)**

(c) the coarse focusing wheel

.... **(1 mark)**

Guided

2 A student viewed plant cells using a light microscope. He made a biological drawing of some of the cells.

Figure 1 shows the image seen through the microscope. **Figure 2** shows the student's drawing.

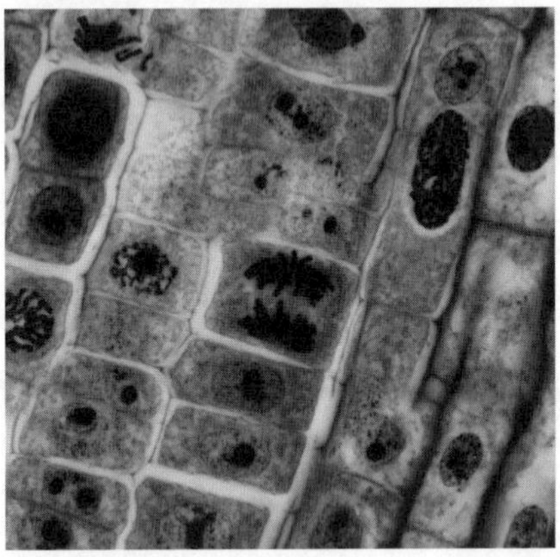

Figure 1

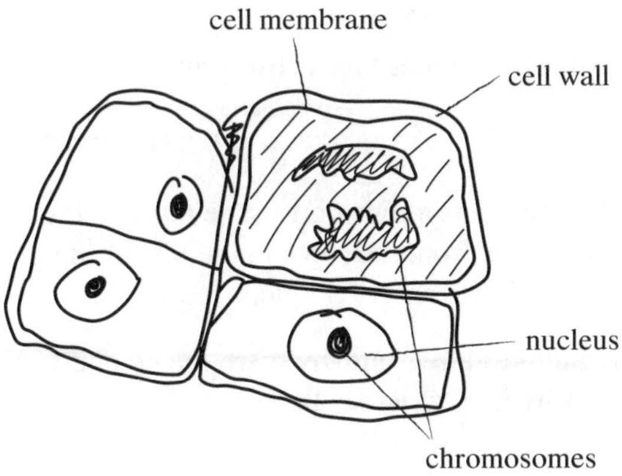

Figure 2

Identify **three** faults with the student's drawing.

1 The drawing is made with a pen rather than with a ....

2 ....

3 .... **(3 marks)**

3 The highest magnification of a microscope allows smaller details to be observed. Describe **two** drawbacks of using the highest magnification rather than the lowest magnification.

....

.... **(2 marks)**

4 A student is observing a slide under high power. She cannot find the part she wants to study. Describe how the student should bring this part into view.

Think about the steps needed to use the microscope safely to make the necessary adjustments.

....

....

.... **(3 marks)**

Aseptic techniques

1 The mean division time for a strain of bacteria is 0.5 hours. A culture with 2000 of these bacteria is incubated. How many bacteria will there be after 1.5 hours? Tick **one** box.

6000 ☐ 8000 ☐ 16 000 ☐ 32 000 ☐ **(1 mark)**

2 Give a reason why the following steps are used when working with cultures of bacteria.

(a) The bench is wiped with disinfectant before starting work.

.... **(1 mark)**

(b) The Petri dish lid is not completely sealed before incubating the culture.

.... **(1 mark)**

(c) In school and college laboratories, cultures are incubated at 25 °C even though they grow faster at 37 °C.

.... **(1 mark)**

(d) The inoculating loop is sterilised after use.

.... **(1 mark)**

3 A student is culturing some bacteria. Here are the steps he uses:

- Agar gel is heated to 80 °C.
- The gel is cooled to about 20 °C and sets solid before adding a sample of bacteria.
- The gel is put into sterilised Petri dishes, then incubated.

Give a reason why each of these steps is important in making the bacterial culture safely and efficiently.

Write about each bullet point in turn.

....

....

....

....

....

.... **(3 marks)**

Guided

4 It is important that people doing experiments to culture bacteria follow some safety precautions. For each precaution given, explain why it is important.

(a) The lid of the Petri dish is opened enough only to inoculate the agar plate.

This will .... bacteria from the air that are

.... **(2 marks)**

(b) The inoculating loop is held in a flame before use.

....

.... **(2 marks)**

Had a go ☐ Nearly there ☐ Nailed it! ☐

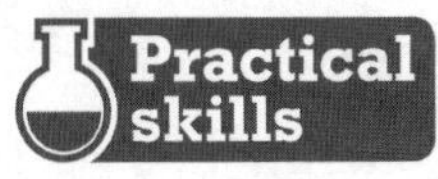

Investigating microbial cultures

1 A student investigated the effect of four different antibiotics on bacterial growth. She placed paper discs containing the antibiotics onto a culture of bacteria in a Petri dish. She then incubated the dish for 2 days. The diagram shows her results.

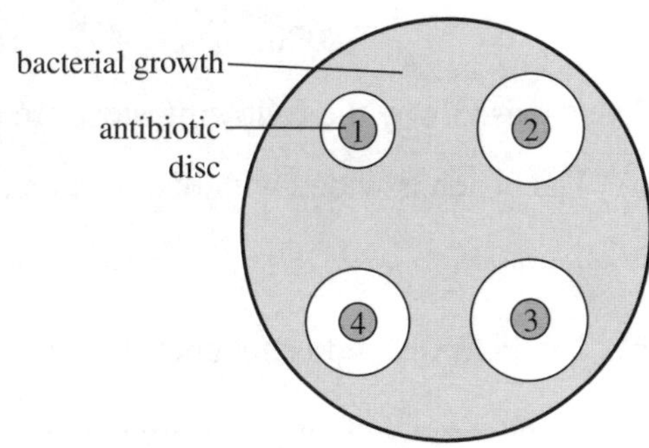

(a) Explain why there are clear zones around each antibiotic disc.

....

.... **(2 marks)**

(b) Complete the table using information in the diagram. Give the areas to 3 significant figures.

Maths skills: Use a ruler to measure the diameter of each clear zone in millimetres. Then calculate the area: area of a circle = $\pi \times r^2$ (where r = radius)

Antibiotic	Diameter of clear zone in mm	Area of clear zone in mm^2
1		
2		
3		
4		

(4 marks)

(c) Identify the most effective antibiotic (**1**, **2**, **3** or **4**). Explain your answer.

....

.... **(2 marks)**

2 (a) The student wanted to investigate the effect of the concentration of one of the antibiotics on bacterial growth. Describe how the student should adapt the method she used in question 1 to do this.

Use different concentrations of ....

.... **(2 marks)**

(b) Give **two** variables the student should control in the investigation.

1 ....

2 .... **(2 marks)**

Mitosis

1 Which of the following are produced when a cell divides by mitosis? Tick **one** box.

two genetically different diploid daughter cells ☐

two genetically identical diploid daughter cells ☐

four genetically identical haploid daughter cells ☐

four genetically different haploid cells ☐ **(1 mark)**

2 Give **three** reasons why mitosis takes place.

Guided

1 to produce new individuals by .... reproduction

2 ....

3 .... **(3 marks)**

3 The photograph shows onion root tip cells viewed through a microscope.

(a) Describe what is happening in the cell labelled **X**.

....

....

....

....

.... **(2 marks)**

(b) Describe what would happen next to the cell labelled **X** so it would form daughter cells.

Think about the cytoplasm and cell membrane of the cell.

....

.... **(2 marks)**

4 The graph shows the mass of DNA in the nucleus of a cell over a 24-hour period.

Give the stages in the cell cycle at the parts on the graph labelled **A**, **B**, **C** and **D**.

Remember that you need to understand the three overall stages of the cell cycle, but do not need to know the different phases of the mitosis stage.

....

....

....

....

....

.... **(4 marks)**

Stem cells

Guided

1 What is a stem cell? Tick **one** box.

an undifferentiated cell ☐

a specialised cell of an organism ☐

a cell found only in embryos ☐

~~a cell that causes diabetes and paralysis~~ ☐

The last option cannot be correct because stem cells may be able to help conditions such as these.

(1 mark)

2 Some plant tissues contain stem cells.

(a) Give the name of the tissue where plant stem cells are found.

.... **(1 mark)**

(b) The tissue named in part (a) is found at the tip of roots.

(i) Give **one** other place in a plant where this tissue is found.

.... **(1 mark)**

(ii) Describe the function of stem cells in the tip of roots.

Include the name of at least one other tissue in your answer.

....

.... **(2 marks)**

Guided

(c) Plant stem cells can be used to produce clones of plants quickly and economically. Describe **two** reasons why people may want to produce such clones.

Rare species can be cloned so they ....

....

.... **(2 marks)**

3 Some disorders may be treated using adult stem cells. For example, leukaemia is a disorder in which white blood cells are produced in excess numbers and do not function normally. Adult stem cells from the bone marrow of a donor are transplanted to the patient, where they differentiate to produce normal white blood cells.

(a) Describe the meaning of the term 'differentiate'.

....

.... **(2 marks)**

(b) Give **two** risks of using adult stem cells for medical treatments.

1 ....

2 .... **(2 marks)**

(c) In therapeutic cloning, an embryo is produced with the same genes as the patient. Suggest **one** advantage and **one** disadvantage of using stems cells from an embryo like this.

Advantage: ....

Disadvantage: .... **(2 marks)**

Diffusion

1 Explain what is meant by the term 'diffusion'.

Guided

Diffusion is the .... of particles, so that there is a net movement of particles from an area of ....

to an area of .... **(2 marks)**

2 Substances can diffuse when they are in the gas state or in solution. The temperature of the gas or solution is one of the factors that affects the rate of diffusion.

(a) Give **three** other factors that can affect the rate of diffusion.

Remember that diffusion can happen across cell membranes.

1 ....

2 ....

3 .... **(3 marks)**

(b) Explain why the rate of diffusion depends upon the temperature.

Think about what happens to the movement of particles in gases and solutions as the temperature increases.

....

.... **(2 marks)**

3 Urea is a waste product. It diffuses from cells into the blood plasma for excretion by the kidneys. Describe how the concentration of urea in these cells compares with its concentration in the blood plasma. Justify your answer.

....

....

.... **(3 marks)**

4 The diagram shows two dissolved substances in neighbouring cells, separated by a cell membrane.

Explain what happens to these two dissolved substances when the cells are left for some time.

....

....

....

.... **(4 marks)**

Had a go ☐ Nearly there ☐ Nailed it! ☐

Exchange surfaces

1 A student investigated how quickly diffusion happens in agar gel. The gel contained dilute sodium hydroxide solution and phenolphthalein indicator, which made it pink. He cut the gel into cubes of different side lengths and placed all the cubes in dilute hydrochloric acid. The student timed how long the cubes took to become completely colourless. Which of the following took the longest time for this change? Tick **one** box.

> Hydrochloric acid diffuses into the cubes and neutralises the sodium hydroxide inside, causing a colour change.

one 5-mm cube ☐ two 10-mm cubes ☐

four 5-mm cubes ☐ one 20-mm cube ☐ **(1 mark)**

2 The small intestine is adapted for the efficient absorption of digested food molecules.

(a) Name the finger-like structures that cover the lining of the small intestine.

.... **(1 mark)**

Guided

(b) Describe **three** ways in which the structures named in part (a) are adapted to provide an effective exchange surface.

Their shape gives them a large ....

They provide a short diffusion path because ....

.... A network of blood capillaries inside them

ensures that .... **(3 marks)**

3 Calculate the surface area to volume ratio of a cube-shaped cell with a side length 50 μm.

> **Maths skills** Calculate the total surface area in μm^2 and the volume in μm^3. Remember that a cube has six equal square sides.

....

....

....

Surface area to volume ratio = ... **(2 marks)**

4 The diagrams show two types of worm, a flatworm and an earthworm. They are similar in size.

Flatworm

Earthworm

The earthworm has a transport system (a heart and blood vessels) but the flatworm does not. Explain these observations.

....

....

....

.... **(4 marks)**

Osmosis

1 Describe what is meant by osmosis.

Guided

Osmosis is the diffusion of .... from a ....

solution to a .... solution through a

.... (4 marks)

2 A student cut two pieces of the same size from a potato. She put one piece of potato into some distilled water. She put the other piece of potato into a strong solution of glucose. She left the potato pieces for 5 hours and then looked to see if they had become longer.

(a) Explain why it is important that the two pieces of potato are the same size at the start.

....

....

.... (2 marks)

(b) Describe what she would notice about the size of each piece of potato at the end of the experiment.

Potato in distilled water: ....

Potato in strong glucose solution: .... (2 marks)

3 This apparatus can be used to model osmosis in cells.

(a) The Visking tubing is partially permeable. Describe what 'partially permeable' means.

....

....

.... (2 marks)

(b) Explain why the level of liquid in the glass tube gradually rises when the apparatus is left for a few hours.

....

....

.... (3 marks)

(c) Predict what would happen to red blood cells placed in water. Explain your answer.

Think about whether the volume of cytoplasm will increase or decrease, and why. What effect will this change have on the appearance of the cells?

....

....

....

.... (4 marks)

Had a go ☐ Nearly there ☐ Nailed it! ☐

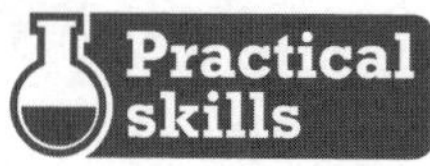

Investigating osmosis

1 A student investigated the effect of solutions with different concentrations of sucrose on the mass of potato tissue. He used a cork borer to cut equal-sized cylinders of potato, then weighed each one. The student placed the cylinders in the different solutions. He removed them after a few hours, dried them with a paper towel, and weighed them again. The table shows his results.

Concentration in mol/dm^3	Initial mass in g	Final mass in g	Change in mass in g	Percentage change in mass
0	2.60	2.85		9.6
0.2	2.51	2.67	0.16	6.4
0.4	2.65	2.72	0.07	2.6
0.6	2.52	2.45	–0.07	–2.8
0.8	2.58	2.43	–0.15	

Guided

(a) Calculate the **two** missing values in the table. Use these values to complete the table.

Change in mass = 2.85 − 2.60 = ... g

Percentage change in mass = (−0.15/2.58) × 100 = ...% **(2 marks)**

(b) Plot a graph to show the percentage change in mass against concentration in mol/dm^3. **(3 marks)**

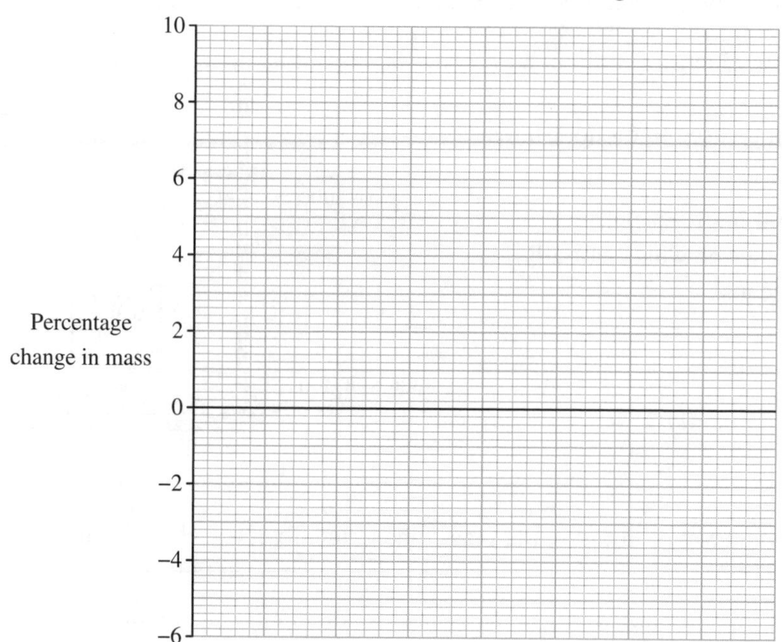

Maths skills Choose a suitable scale for the horizontal axis. The scale should allow the plotted points to cover at least half of the area of the graph.

Draw a line of best fit. This can be curved or straight, depending on the data, but should ignore points that are clearly anomalies.

(c) Use the completed graph to estimate the concentration of the potato tissue.

.... **(1 mark)**

(d) The student used a balance with a resolution of ±0.01 g. Explain, using the readings at 0.2 mol/dm^3 as an example, why the student did not use a balance with a resolution of ±0.1 g instead.

The **resolution** of an instrument is the smallest change in the quantity being measured that gives a perceptible change in the reading. What readings would a ±0.1 g balance give?

....

....

.... **(3 marks)**

Active transport

1 Cells move sodium ions from a low concentration inside the cell to a high concentration outside the cell. What process do cells use to do this? Tick **one** box.

diffusion ☐ osmosis ☐ active transport ☐ dissolving ☐ **(1 mark)**

2 Plants require nitrate ions for healthy growth. Plants move nitrate ions from very dilute solutions in the soil to higher concentrations in the root hair cells.

(a) Give a reason that explains why nitrate ions cannot be moved by osmosis.

.... **(1 mark)**

(b) Use the information given to explain why diffusion is not responsible for moving these ions.

Think about what happens during diffusion.

....

.... **(2 marks)**

Guided

(c) Explain why the rate of respiration may increase in root hair cells during the uptake of nitrate ions.

The nitrate ions are being moved by ....

This process requires .... from .... **(3 marks)**

3 Active transport is used to move dissolved glucose from the gut to the blood plasma.

(a) Describe an advantage of absorbing glucose into the blood in this way.

....

.... **(1 mark)**

(b) Some toxins prevent the release of energy by mitochondria. Predict the effect of these toxins on the absorption of glucose into the plasma. Explain your answer.

....

....

.... **(3 marks)**

4 Substances are transported into and out of cells by diffusion, osmosis and active transport.

(a) Compare the main features of diffusion and osmosis.

Describe the similarities and differences between the two processes.

....

.... **(2 marks)**

(b) Compare the main features of diffusion and active transport.

....

....

.... **(3 marks)**

Had a go ☐ Nearly there ☐ Nailed it! ☐

Extended response – Cell biology

The diagrams show a bacterial cell and a plant cell. The diagrams are not drawn to scale.

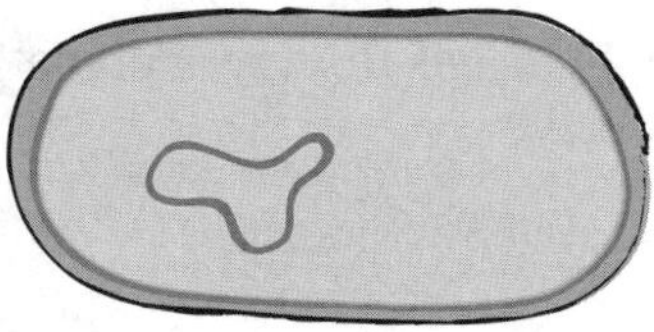

Bacterial cell

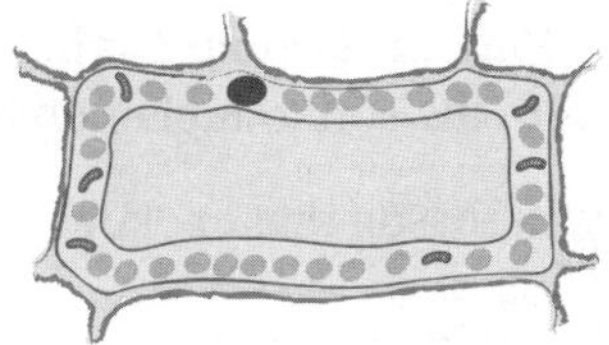

Plant cell

Compare the structures of these two cells, including sub-cellular structures and their functions.

In your answer to this question, you need to think about:

- the similarities between the cells
- the differences between the cells.

For each structure that you identify, remember to describe its function.

It may help if you make a brief plan before you start writing.

....

....

....

....

....

....

....

....

....

....

....

....

....

....

....

....

....

.... **(6 marks)**

The digestive system

1 The diagram shows part of the human digestive system. Identify the organs labelled A to E.

Organs in the digestive system include the liver, large intestine, pancreas, small intestine and stomach.

A ....

B ....

C ....

D ....

E .... **(5 marks)**

2 The stomach is an organ in the digestive system. It is composed of several types of tissue.

Guided

(a) Describe what is meant by a tissue.

A tissue is a group of cells with a similar .... and

.... **(2 marks)**

(b) Describe what is meant by an organ.

....

.... **(2 marks)**

(c) The digestive system is an organ system. Name **two** other organ systems in the human body.

1 ....

2 .... **(2 marks)**

3 Digestive enzymes in the digestive system convert large, insoluble molecules in food into small soluble molecules.

(a) Complete the table to show the features of carbohydrase, protease and lipase enzymes.

The **substrate** is the substance changed by an enzyme.

Type of enzyme	Substrate	Product(s)
carbohydrase		simple sugars
protease	proteins	
lipase		

(4 marks)

(b) Amylase is an example of a carbohydrase. Name the substance broken down by amylase.

.... **(1 mark)**

Had a go ☐ Nearly there ☐ Nailed it! ☐

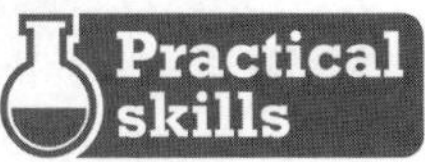

Food testing

1 A student carried out a test to detect lipids in a food sample. This is the method she used.

Detecting lipids	
A	Grind up a small sample of dry food and transfer it to a beaker.
B	Add distilled water and stir to disperse the food.
C	Half fill a test tube with this mixture and add three drops of Sudan III stain.
D	Shake gently to mix, and record your observations.

(a) Name suitable laboratory apparatus that can be used to grind up dry food samples.

.... **(1 mark)**

(b) What will the student observe if the food sample contains lipids? Tick **one** box.

a blue-stained layer floating on a layer of water ☐

a blue-stained layer underneath a layer of water ☐

a red-stained layer floating on a layer of water ☐

a red-stained layer underneath a layer of water ☐ **(1 mark)**

Guided

2 Describe the test you would use to find out if protein is present in egg white.

Put some egg white in a test tube. Add an equal volume of ....

.... and shake to mix. If protein is present, the mixture turns

.... **(2 marks)**

3 A student carried out a test on samples of two different foods. He dissolved each sample in water and added Benedict's solution. The student heated the mixtures in test tubes for about 5 minutes, and then recorded his observations.

(a) Describe how the student can heat the mixtures safely.

> A Bunsen burner is not necessary to carry out these tests.

....

.... **(2 marks)**

(b) One mixture turns green and the other turns red. Explain what these observations show.

....

.... **(2 marks)**

4 Flour is a powdery dry food.

(a) Describe the test you would use to find out if starch is present in a sample of flour.

> Say what you would do and what you would see.

....

.... **(2 marks)**

(b) Give **one** hazard associated with the reagent used in this test.

.... **(1 mark)**

Enzymes

Guided

1 'Biological' washing powders contain enzymes including proteases and lipases. These break down food stains on clothes. Explain why proteases can break down proteins in food stains but lipases cannot.

The active site in proteases matches the shape of .... but

.... **(2 marks)**

2 Certain bacteria are adapted to live in hot water springs. The graph shows how the activity of an enzyme found in these bacteria is affected by temperature.

(a) Give the optimum temperature for this enzyme.

.... **(1 mark)**

(b) Explain why enzyme activity increases between 10 °C and 50 °C.

Think about the rate of collisions involving molecules.

....

.... **(2 marks)**

(c) Explain the change in enzyme activity above 70 °C.

Use your knowledge of the effect of high temperatures on the structure of proteins such as enzymes.

....

....

....

.... **(3 marks)**

3 Pepsin and trypsin are proteases. Pepsin is produced in the stomach (pH 2) and trypsin is found in pancreatic juice (pH 8.6) released into the small intestine. Saliva (pH 7.5) and pancreatic juice both contain amylase. The graph shows the effect of pH on the activity of these three enzymes.

Proteins are digested in the stomach and small intestine, but starch is digested only in the mouth and small intestine. Use the information to explain why.

....

....

....

.... **(4 marks)**

Had a go ☐ Nearly there ☐ Nailed it! ☐

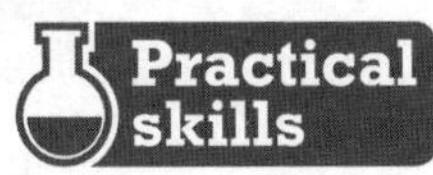

Investigating enzymes

1 A student investigated the effect of pH on the activity of trypsin. Trypsin digests the proteins in photographic film, turning it clear. The student used the apparatus shown in the diagram. She measured the time taken for trypsin solution to turn pieces of film clear at different pH values.

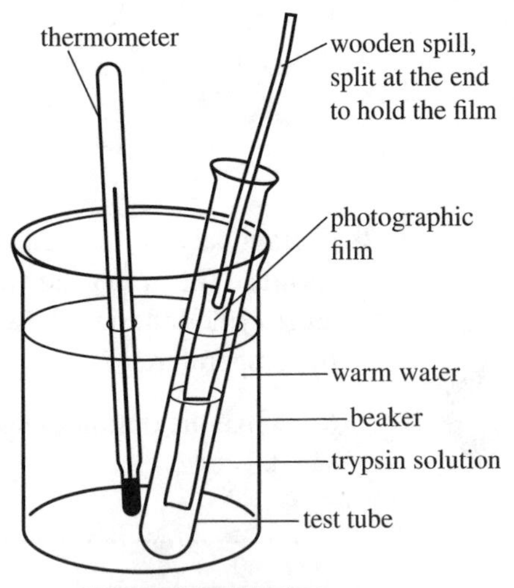

The table shows her results.

pH	2	4	6	8	10
Time in min	>10	7.5	3.6	1.2	8.3
Rate in /min	0	0.13			

(a) Complete the table by calculating the rate at each pH. **(2 marks)**

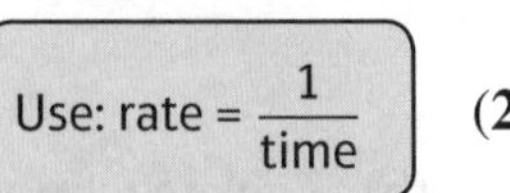

(b) Plot a graph to show the rate of reaction against pH. **(4 marks)**

Choose scales that allow the plotted points to cover at least half the area of the graph. Remember to label both axes and draw a line of best fit.

(c) Describe **two** improvements the student could make to her method.

1 ....

2 .... **(2 marks)**

The blood

1 Draw **one** line from each blood component to a correct function.

Blood component	Function
plasma	carries other blood components
platelet	part of the body's immune system
red blood cell	involved in forming blood clots
white blood cell	carries oxygen

(4 marks)

2 Blood contains red blood cells.

(a) Name the cell structure, normally found in cells, that is missing in human red blood cells.

.... **(1 mark)**

(b) Name the compound in red blood cells that gives them their colour.

.... **(1 mark)**

Guided

(c) The diagram shows some red blood cells.

Describe **two** ways in which red blood cells are adapted to carry out their function.

Their biconcave shape gives them a large ....

for diffusion to happen efficiently. They are also flexible, which lets them ....

.... **(2 marks)**

3 The plasma transports soluble products of digestion, including glucose and amino acids. Name **two** waste substances transported by the plasma.

1 ....

2 .... **(2 marks)**

4 There are different types of white blood cells, phagocytes and lymphocytes. Describe a function of each type of cell.

Phagocytes are named after the Greek word 'phagein', which means 'to eat', but do not write that phagocytes *eat* pathogens (disease-causing organisms).

Phagocyte: ....

.... **(1 mark)**

Lymphocyte: ....

.... **(2 marks)**

5 Explain the role of platelets in protecting the body from infection.

....

....

.... **(2 marks)**

Blood vessels

1 Which of the following types of blood vessel contains blood at the lowest pressure? Tick **one** box.

artery ☐ vein ☐ capillary ☐ **(1 mark)**

2 The diagram shows a cross-section through a vein and an artery.

(a) Give a reason why the vein has a large lumen.

.... **(1 mark)**

(b) Explain why the artery has a thick wall with muscle and elastic fibres.

> Think about the reason why the artery wall is thick, and why it contains muscle and elastic fibres.

....

....

....

.... **(3 marks)**

3 Substances diffuse between blood in the capillaries and the body cells around them. Explain how the capillaries are adapted to this function.

Guided

The capillaries are about as wide as one red blood cell, so the distance oxygen must travel to the capillary wall is .... The walls are only one cell thick, so .... **(2 marks)**

4 Veins carry blood away from the organs of the body to the heart.

(a) Explain how the structure of veins helps them carry blood back to the heart.

....

....

.... **(2 marks)**

(b) A nurse taking a sample of blood from a patient will insert a needle into a vein. Suggest a reason that explains why the nurse does not take blood from an artery instead.

....

....

.... **(2 marks)**

The heart

1 The heart is connected to four major blood vessels. Complete the table below.

Guided

Blood vessel	Carries blood from	Carries blood to	Carries oxygenated blood (✓ or ✗)
aorta	heart	body	✓
pulmonary artery			
pulmonary vein			
vena cava			

(4 marks)

2 (a) Explain why the heart is mainly muscle.

....

.... **(2 marks)**

(b) Describe the route taken by blood through the heart from the vena cava to the aorta.

....

....

.... **(3 marks)**

3 The diagram shows a section through the human heart.

Remember that the heart is drawn and labelled as if you are looking towards the front of someone's chest. So the left side of the heart is shown on the right of the diagram.

(a) Explain the function of the part labelled **A**.

....

....

.... **(2 marks)**

(b) Name the part labelled **B** and describe its function.

....

.... **(2 marks)**

(c) Explain why the muscle at **C** must be thicker than the muscle on the other side of the heart.

....

....

.... **(3 marks)**

The lungs

1 The diagram shows parts of the respiratory system.

Identify the structures labelled **A** and **B**.

A ....

B .... **(2 marks)**

A
B
Lung

2 Gas exchange happens at the surface of the alveoli in the lungs.

(a) Name the process by which gas exchange happens.

.... **(1 mark)**

(b) Describe the directions in which gas exchange happens.

Guided

There is a net movement of carbon dioxide from..

to ...., and a net movement of oxygen from

.... **(2 marks)**

(c) Explain **two** ways in which the structure of the alveoli is adapted for efficient gas exchange.

....

....

....

.... **(4 marks)**

3 Emphysema is a type of lung disease where elastic tissue in the alveoli breaks down. The diagram shows the appearance of an alveolus damaged by lung disease compared with a healthy alveolus.

Suggest a likely symptom of emphysema and explain your answer.

> Think about the changes shown in the diagram and how they might affect lung function.

Healthy alveolus

Alveolus damaged by lung disease

....

....

.... **(3 marks)**

Cardiovascular disease

1 In coronary heart disease, layers of fatty material build up inside the coronary arteries.

(a) Explain how this can lead to a heart attack.

Which organ is supplied with blood by the coronary arteries?

....

....

.... **(3 marks)**

(b) Describe how coronary heart disease may be treated using:

Say what each treatment is and what it does.

(i) a stent

....

.... **(2 marks)**

(ii) statins

....

.... **(2 marks)**

2 In some people, the heart valves may become faulty.

(a) Give **two** ways in which a heart valve may not function properly.

1 ....

2 .... **(2 marks)**

(b) Give **two** ways in which faulty heart valves can be replaced.

1 ....

2 .... **(2 marks)**

(c) Give **one** problem caused by faulty heart valves.

....

.... **(1 mark)**

Guided

3 The table summarises some of the benefits and drawbacks of the different types of treatment for cardiovascular disease.

Type of treatment	Benefits	Drawbacks
lifestyle changes such as dietary change	no side effects, may reduce risk of other health problems	may take a long time to work, may not work effectively
medication with drugs		
surgery, including transplants		

Complete the table to show a benefit and drawback of each type of treatment. **(6 marks)**

Had a go ☐ Nearly there ☐ Nailed it! ☐

Health and disease

1 Health may be described as the state of physical and mental wellbeing.

Disease is a factor that can affect health. Give **one** other factor that can affect health.

What might affect physical wellbeing or mental wellbeing?

.... **(1 mark)**

2 Diseases can be communicable or non-communicable.

(a) Name the type of disease that can be passed from person to person.

.... **(1 mark)**

(b) Give **one** example of:

(i) a communicable disease

.... **(1 mark)**

(ii) a non-communicable disease.

.... **(1 mark)**

Guided

(c) Complete the table to show features of these two types of disease.

	Communicable disease	Non-communicable disease
Number of cases	rapid variation over time	
Distribution of cases		

(2 marks)

3 A student was interested in studying how diseases may interact with each other. This is part of the student's research notes.

> The scabies mite is a tiny arthropod. Holding hands with an infected person for a lengthy time can transmit it. The mite burrows into the skin and lays its eggs there. Burrow marks usually appear in warm places, such as skinfolds. However, a lumpy red rash can appear anywhere on the body. This allergic immune response causes severe itching, and repeated scratching may break the skin's surface.

Use this information to help you answer these questions.

(a) Explain why scabies is a communicable disease.

....

.... **(2 marks)**

(b) Explain why itchy skin itself is a non-communicable disease.

....

.... **(2 marks)**

(c) People infected with the scabies mite are more likely to develop a bacterial skin infection. Suggest a reason to explain why.

....

.... **(2 marks)**

Lifestyle and disease

1 The chart shows how the body type of a person is related to their mass and height.

Suggest the advice that a doctor might give to a person who is 190 cm tall and has a mass of 120 kg. Explain your answer.

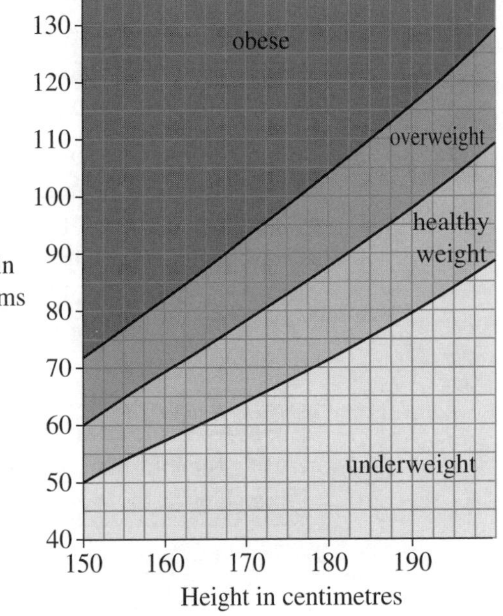

Use the chart to determine the person's body type, then consider how they could achieve a healthy weight.

....

....

....

.... **(3 marks)**

2 Ionising radiation is a risk factor for developing cancer.

(a) Give **two** features of a risk factor with a proven causal mechanism.

1 ....

2 .... **(2 marks)**

(b) Describe what is meant by cancer.

....

.... **(2 marks)**

3 The chart shows the results of a survey into the relationship between body mass and the incidence of Type 2 diabetes.

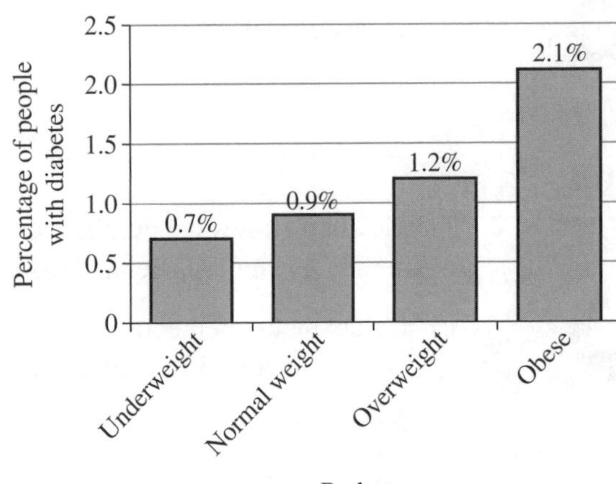

(a) Describe the trend shown by this graph.

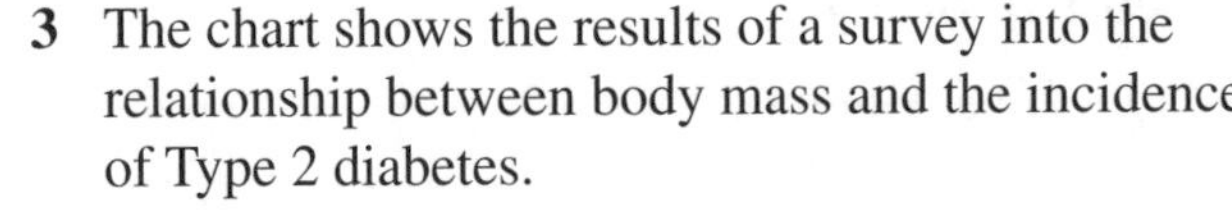

....

.... **(2 marks)**

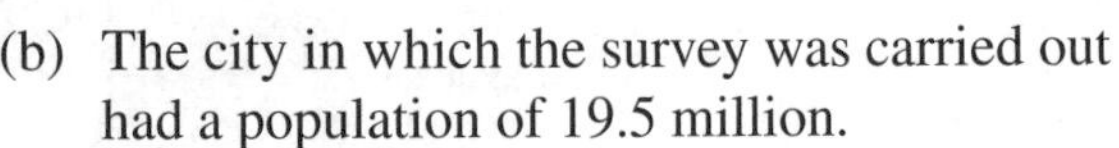

Guided

(b) The city in which the survey was carried out had a population of 19.5 million.

(i) Calculate the number of people in that city who had Type 2 diabetes.

Total percentage of people with diabetes = ...

Number of people with diabetes = ... ÷ 100 × ....

= .... **(3 marks)**

(ii) Give **one** reason why the number calculated in part (i) may not be valid.

.... **(1 mark)**

Had a go ☐ Nearly there ☐ Nailed it! ☐

Alcohol and smoking

1 The graph shows how the risk of having a car accident changes at different concentrations of alcohol in the blood.

(a) Each glass of wine increases John's blood alcohol concentration by 20 mg per 100 cm^3 of blood. Use the graph to determine John's increased risk of having an accident if he drinks five glasses of wine.

Maths skills Calculate John's blood alcohol concentration using the information given, and then look at the graph.

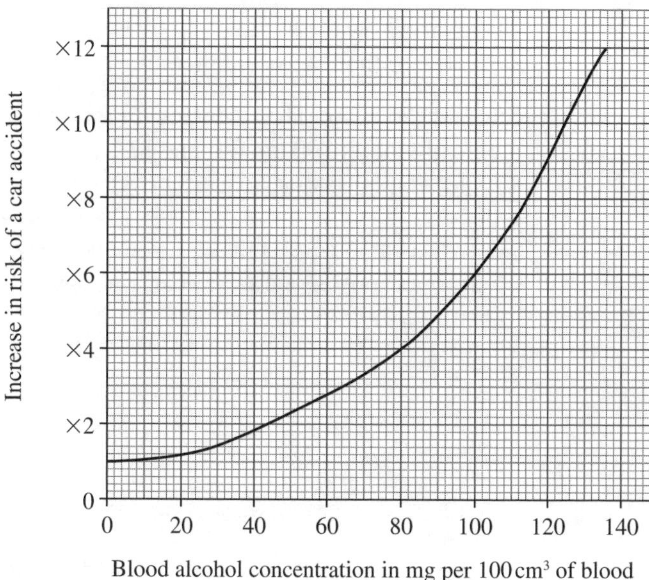

....

....

Increased risk of having an accident.. **(2 marks)**

(b) Describe the trend shown by the graph.

Guided

As the concentration of alcohol in the blood rises, the risk of having a car accident .... *The risk of having an accident increases as the* .... **(2 marks)**

(c) Give a reason that explains the trend shown by the graph.

....

.... **(1 mark)**

2 The pie chart shows the number of deaths caused each year in America by smoking-related diseases.

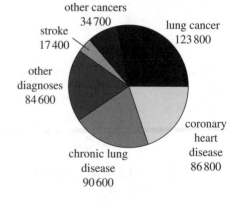

other cancers 34 700
stroke 17 400
lung cancer 123 800
other diagnoses 84 600
coronary heart disease 86 800
chronic lung disease 90 600

(a) Calculate the percentage of deaths from lung cancer. Give your answer to 3 significant figures.

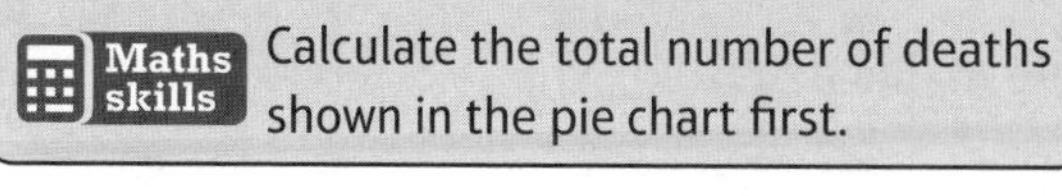

Maths skills
Calculate the total number of deaths shown in the pie chart first.

....

....

Percentage of deaths from lung cancer .... **(2 marks)**

(b) Other than the risk of developing the diseases shown in the pie chart, explain why pregnant women are advised not to smoke.

....

.... **(2 marks)**

The leaf

1 Which of the following best describes a leaf? Tick **one** box.

cell ☐ tissue ☐ organ ☐ organ system ☐ **(1 mark)**

2 Leaves are supported by the stem, which is connected to the root. Give **two** functions of the root.

1 ....

2 .... **(2 marks)**

3 The diagram shows a cross-section of part of the leaf of a plant.

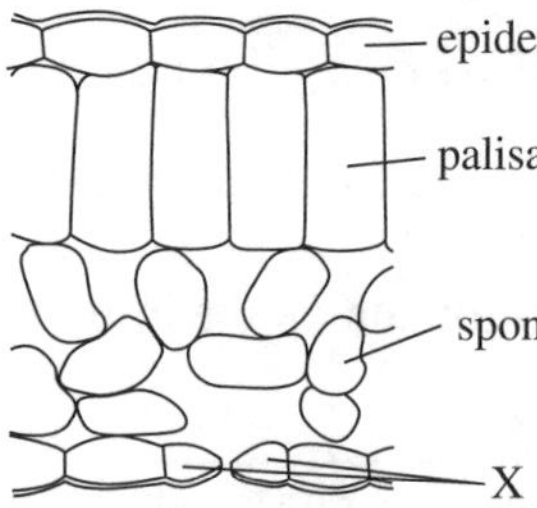

epidermal cells
palisade mesophyll cells
spongy mesophyll cell
X

(a) The cells labelled **X** are found mostly on the underside of the leaf.

(i) Identify these cells.

.... **(1 mark)**

(ii) Name the structure formed by these cells.

.... **(1 mark)**

(b) The upper epidermis is covered by a layer called the cuticle. Suggest a reason why:

Guided

(i) the cuticle is thin and transparent

This is so that light can .... **(1 mark)**

(ii) the cuticle is waxy

.... **(1 mark)**

(c) Explain how the following are adapted for photosynthesis:

(i) palisade mesophyll tissue

Think about the shape of the cells involved and relevant sub-cellular structures the cells may contain.

....

....

.... **(2 marks)**

(ii) spongy mesophyll tissue

Notice how there are air spaces in this layer.

....

.... **(2 marks)**

Had a go ☐ Nearly there ☐ Nailed it! ☐

Transpiration

1 How do mineral ions and water enter root hair cells? Tick **one** box.

Mineral ions and water enter by active transport. ☐

Mineral ions enter by diffusion and water by active transport. ☐

Mineral ions enter by active transport and water by osmosis. ☐

Mineral ions enter by diffusion and water by osmosis. ☐ **(1 mark)**

2 Describe the process of transpiration.

Guided

Water .... from the leaves, mostly through the

.... This causes a pull so that water moves through

the .... and is replaced by water entering the roots. **(3 marks)**

3 A student investigated the number of stomata on the upper and lower surfaces of a leaf. This is the method she used.

Coat the leaf surface with colourless nail varnish and let it dry.

Peel off the dry layer of nail varnish with sticky tape and stick it onto a microscope slide.

Observe the slide using a light microscope and count the stomata in several equal-sized areas.

The table shows the student's results.

Area	Number of stomata	
	Lower surface	Upper surface
A	22	0
B	18	2
C	23	3
D	19	1

Maths skills: To find the mean, add all the values together then divide the total by the number of values.

(a) Calculate the mean number of stomata on each surface.

Lower surface: ....

Upper surface: .... **(2 marks)**

(b) Predict which surface should lose most water, and explain your answer.

....

.... **(2 marks)**

(c) The size of the opening in a stoma can vary, depending on the external conditions.

(i) Give **one** advantage to a plant of having closed stomata when the soil is dry.

.... **(1 mark)**

(ii) Explain why it is important for a plant that its stomata do not remain completely closed.

....

....

.... **(2 marks)**

Investigating transpiration

1 The apparatus in the diagram was used to investigate the movement of water in the shoot of a plant. Circle A shows the position of an air bubble at the start of the experiment. Circle B shows its position at the end.

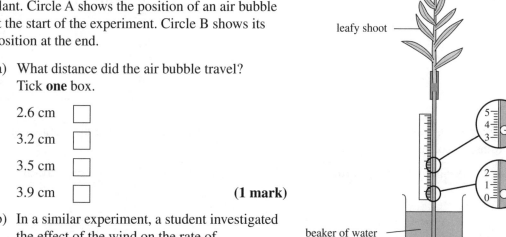

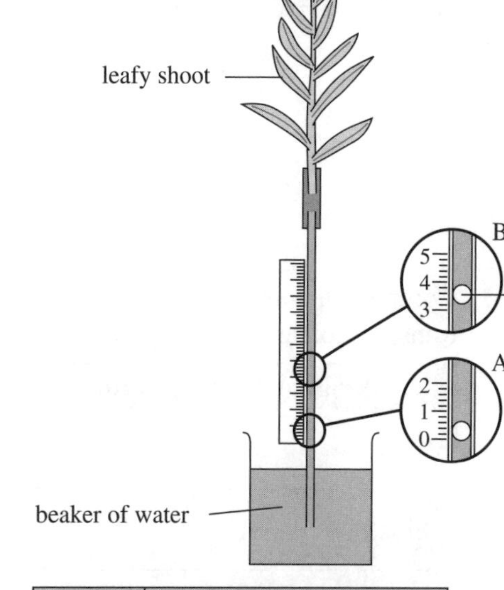

(a) What distance did the air bubble travel? Tick **one** box.

2.6 cm ☐

3.2 cm ☐

3.5 cm ☐

3.9 cm ☐ **(1 mark)**

(b) In a similar experiment, a student investigated the effect of the wind on the rate of transpiration. He used an electric fan to simulate wind blowing. He measured the distance travelled by the bubble in 5 minutes with the fan off or on. The table shows the student's results.

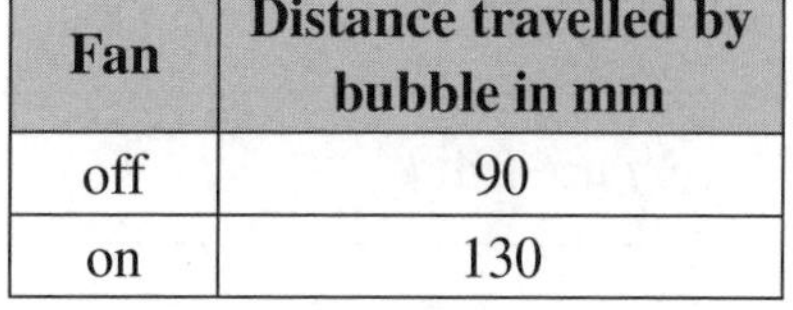

Fan	Distance travelled by bubble in mm
off	90
on	130

Guided

(i) The bubble travelled through a capillary tube with a diameter of 0.5 mm. Calculate the rate of transpiration in mm^3/min when the fan was off. Give your answer to 1 decimal place.

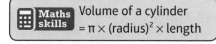

Maths skills: Volume of a cylinder $= \pi \times (\text{radius})^2 \times \text{length}$

radius of tube $= \frac{0.5}{2} =$ .... mm

volume travelled $= \pi \times$ .... $\times 90 =$ .... mm^3

rate of transpiration $= \frac{\text{..............................}}{5} =$ .... mm^3/min **(2 marks)**

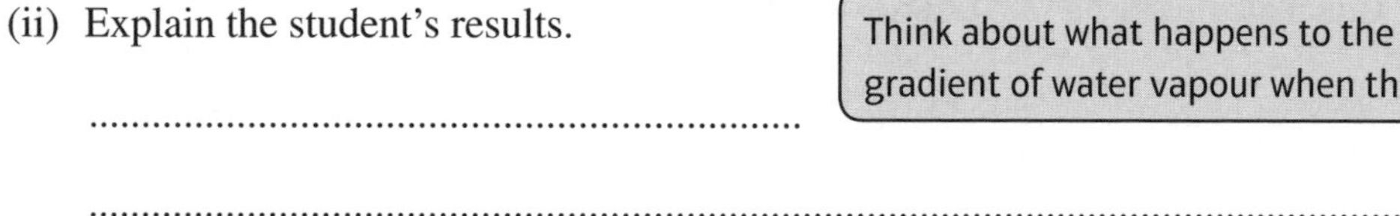

(ii) Explain the student's results.

Think about what happens to the concentration gradient of water vapour when the fan is on.

....

....

....

.... **(3 marks)**

2 (a) Explain why increasing the light intensity may increase the rate of transpiration.

Think about what happens to the stomata.

....

.... **(2 marks)**

(b) Explain why increasing the temperature may increase the rate of transpiration.

....

....

.... **(3 marks)**

Had a go ☐ Nearly there ☐ Nailed it! ☐

Translocation

1 In which plant tissue does translocation take place? Tick **one** box.

phloem ☐ xylem ☐ meristem ☐ mesophyll ☐ **(1 mark)**

2 Describe what is meant by translocation.

....

.... **(2 marks)**

Guided

3 Complete the table by placing a tick (✓) in each correct box to compare the features of transpiration and translocation.

Structure or mechanism	Transpiration	Translocation
xylem	✓	
phloem		
pulled by evaporation from the leaf		
energy needed		

(4 marks)

4 The rate of translocation in a growing plant was measured using the method described below.

Enclose the leaves with a plastic bag and seal the bag against the stem.
Add carbon dioxide containing radioactive carbon atoms to the plastic bag.
Extract sucrose solution from two different places in the stem at various times.
Analyse the sucrose solution for the presence of radioactivity.

(a) Sucrose solution was extracted from two places 0.4 m apart. The time taken for radioactivity to travel between these places was 67 minutes. Calculate the mean rate of translocation in mm/s.

Maths skills: Remember that 1 m = 1000 mm, and 1 minute = 60 s.

....

....

Mean rate = ... **(2 marks)**

(b) Explain why radioactive carbon atoms become part of sucrose molecules in the experiment.

Think about what happens during photosynthesis.

....

....

.... **(3 marks)**

5 Meristem tissue is found at the growing tips of shoots and roots. Describe the function of the stem cells in this tissue.

....

.... **(2 marks)**

Extended response – Organisation

The diagram shows the main features of the human heart and circulatory system.

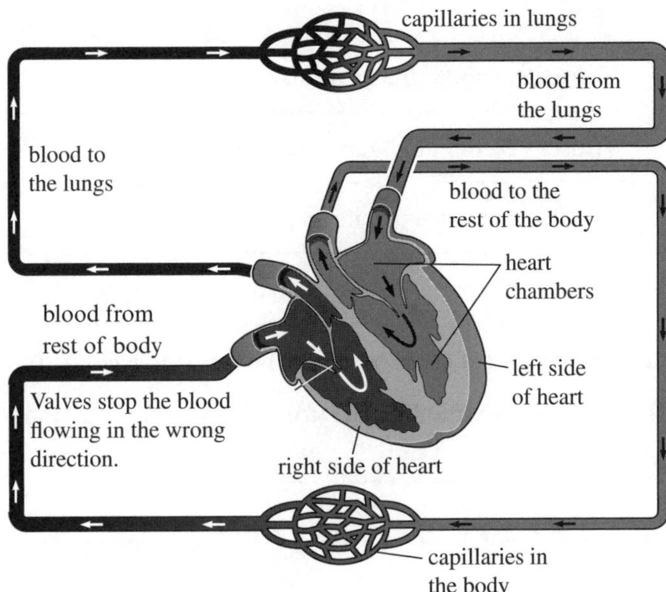

Describe the journey taken by blood around the body and through the heart, starting from when it enters the right side of the heart. In your answer, include the names of major blood vessels and chambers in the heart.

It may help if you make a brief plan before you start writing. You could follow the blood around the diagram with a finger, writing the name of each blood vessel or chamber in order as you go.

To answer this question, you do **not** need to:

- explain how the different components of the heart and circulatory system work, or
- identify any blood vessels in the 'rest of the body' other than the aorta.

....

....

....

....

....

....

....

....

....

....

....

.... **(6 marks)**

Had a go ☐ Nearly there ☐ Nailed it! ☐

Viral diseases

1 Which of the following is a disease caused by a virus? Tick **one** box.

rose black spot ☐ gonorrhoea ☐ malaria ☐ influenza ☐ **(1 mark)**

2 The table shows some possible features of viruses. Some of these are correct and some are incorrect.

Feature	Incorrect (✗)
Viruses are about the same size as cells.	
Viruses can infect plants or animals.	
Viruses reproduce outside cells.	
Viruses are spread by direct contact, air or water.	

(a) Complete the table by placing a cross (✗) in the box next to each incorrect feature. **(2 marks)**

(b) For each incorrect feature identified in part (a), write a correct version of that feature.

....

.... **(2 marks)**

3 Tobacco mosaic virus (TMV) is a plant pathogen. It destroys chloroplasts in the leaves.

Guided

(a) Infection with TMV causes a distinctive 'mosaic' pattern of lighter-coloured areas on the leaves. Suggest an explanation for why parts of the leaves become discoloured.

Chloroplasts contain .... Lighter-coloured areas appear

where .... **(2 marks)**

(b) Explain why infected plants show stunted growth.

....

....

.... **(3 marks)**

4 Measles is a viral disease that causes a fever.

(a) Give **one** other symptom of measles.

.... **(1 mark)**

(b) Give a reason why most young children are vaccinated against measles.

What can happen if complications occur during a measles infection?

.... **(1 mark)**

(c) Describe **one** way in which the measles virus can be spread from person to person.

.... **(1 mark)**

5 HIV is a virus that can infect humans. Explain why other pathogens may more easily infect people who have late-stage HIV infection.

....

.... **(2 marks)**

Bacterial diseases

1 Which of the following is a disease caused by a bacterium? Tick **one** box.

TMV ☐

cold ☐

cholera ☐

ringworm ☐ **(1 mark)**

2 Salmonella food poisoning is spread by bacteria in or on food.

Give **two** symptoms of salmonella food poisoning.

1 ....

2 .... **(2 marks)**

3 Joseph Lister was a surgeon in the nineteenth century. In those days, surgeons did not wash their hands before operations. Patients often died from bacterial infections after surgery.

(a) Suggest a reason that explains why patients became infected during surgery.

.... **(1 mark)**

(b) Lister advised surgeons to use a fine spray of phenol solution during surgery. This substance had been used to clean sewers at the time. The table shows the effect of following Lister's advice.

Number of patients dying from infections per 100 operations	
Without Lister's advice	**With Lister's advice**
45	15

Describe the effect of Lister's advice on the number of patients dying from infections.

Use information from the table in your answer.

....

.... **(2 marks)**

4 A patient has a sore throat caused by a bacterial infection. Explain how this infection makes the patient ill.

Guided

Bacteria get into the body, where they ....

The bacteria produce .... which damage .... **(2 marks)**

5 Gonorrhoea is a sexually transmitted disease (STD) caused by a bacterium.

(a) Give **one** symptom of gonorrhoea.

.... **(1 mark)**

(b) In the past, infection was easily treated with penicillin. This treatment is much less successful today. Explain these observations.

....

....

.... **(3 marks)**

Had a go ☐ Nearly there ☐ Nailed it! ☐

Fungal and protist diseases

1 Different pathogens cause different diseases. Which row in the table shows the type of pathogen that causes each disease? Tick **one** box.

Rose black spot	Malaria	
fungus	protist	☐
fungus	fungus	☐
protist	fungus	☐
protist	protist	☐

(1 mark)

2 Athlete's foot is a fungal disease. It commonly affects the skin between the toes, causing itchy, sore and flaky skin.

(a) Name the type of substance that can be used to treat fungal infections.

You do not need to identify a particular cream or powder in your answer.

.... **(1 mark)**

(b) Suggest why you are more likely to get athlete's foot if you share towels with other people.

....

.... **(2 marks)**

Guided

3 Rose black spot is a disease of plants. Purple or black spots develop on the leaves. These leaves often turn yellow and drop early. Explain why rose black spot affects the growth of infected plants.

If the leaves are damaged or there are fewer leaves, ...

is reduced. The plant cannot make enough .... **(2 marks)**

4 A microorganism called *Plasmodium* causes malaria, a disease that can be fatal. *Plasmodium* has a life cycle that includes the mosquito.

(a) Give **one** symptom of malaria.

.... **(1 mark)**

(b) Describe the role of the mosquito in the spread of malaria.

....

.... **(2 marks)**

(c) Explain why the use of mosquito nets helps to reduce the risk of infection with *Plasmodium*.

....

....

.... **(3 marks)**

Human defence systems

1 The human body has some non-specific defence mechanisms against pathogens.

(a) Give **one** way in which the following features defend the body against pathogens.

(i) Hairs in the nose.

.... **(1 mark)**

(ii) Hydrochloric acid in the stomach.

.... **(1 mark)**

(b) Describe the role of the skin in protecting the body from infection.

....

.... **(2 marks)**

(c) Tears contain an enzyme that helps to protect the body from infection.

(i) Name the enzyme.

.... **(1 mark)**

(ii) Describe how this enzyme protects the eyes against infection.

....

.... **(2 marks)**

2 The diagram shows a section of epithelium in a human bronchiole, one of the tubes in the lungs.

(a) Name substance **A**.

.... **(1 mark)**

(b) Describe the role of substance **A** in protecting the lungs from infection.

Give a feature of this substance and say what it does to pathogens.

....

.... **(2 marks)**

(c) The structure labelled **B** is a part of the epithelial cells.

(i) Name structure **B**.

.... **(1 mark)**

Guided

(ii) Describe how these epithelial cells help to protect the lungs from infection.

The structures on the surface of cells move in ...., which

move ....

.... **(3 marks)**

Had a go ☐ Nearly there ☐ Nailed it! ☐

The immune system

Guided

1 White blood cells are an important part of the immune system. Some of these cells produce antibodies.

(a) Describe how antibodies defend the body against pathogens.

What type of substance are antibodies, and what do they do?

Antibodies are .... They attach to ....

produced by the pathogen, which leads to its destruction. **(2 marks)**

(b) Describe **two** other ways in which white blood cells help to defend against pathogens.

1 ....

....

2 ....

....

.... **(4 marks)**

2 The graph shows the concentration of antibodies in the blood of a child. The lines labelled **A** show the concentration of antibodies effective against the measles virus. The line labelled **B** shows the concentration of antibodies effective against the chickenpox virus.

Concentration of antibodies in blood

A A B 1 2

0 1 2 3 4 5 26 27 28 29 30 31

Weeks

(a) At the time shown by arrow **1**, there was an outbreak of measles. The child had not been exposed to the measles virus before. Explain the shape of line **A** in the 5 weeks after arrow **1**.

....

....

.... **(4 marks)**

(b) During an outbreak of measles and chickenpox 5 months later (arrow **2**), the child was exposed to both viruses. Explain the shape of line A in the first week after arrow 2.

....

....

.... **(3 marks)**

(c) Explain whether the child had been exposed to the chickenpox virus in the past.

Use the information in the graph and your knowledge and understanding of the body's immune response.

....

.... **(2 marks)**

(d) In the second outbreak of measles, the child showed no symptom of measles. Explain why.

....

.... **(2 marks)**

Vaccination

1 Vaccination can prevent illness in an individual and reduce the spread of a pathogen in a population.

(a) Describe what a vaccine contains.

....

.... **(1 mark)**

(b) Give **two** drawbacks of vaccination.

Include a general side effect that some people may develop temporarily.

....

.... **(2 marks)**

Guided

(c) Vaccination prevents a person from becoming ill from infection with a pathogen. This works even if they are exposed to the pathogen a long time after receiving the vaccine. Explain why.

The vaccine causes white blood cells to make .... against the pathogen. If the same pathogen enters the body again, white blood cells respond ....

.... **(3 marks)**

2 The MMR vaccine is effective against measles, mumps and rubella. In 1998, a group of doctors suggested that there was a connection between receiving the MMR vaccine and developing autism. This made some parents afraid of having their babies vaccinated. The graph shows how the percentage of babies in the UK who were given the MMR vaccine changed afterwards.

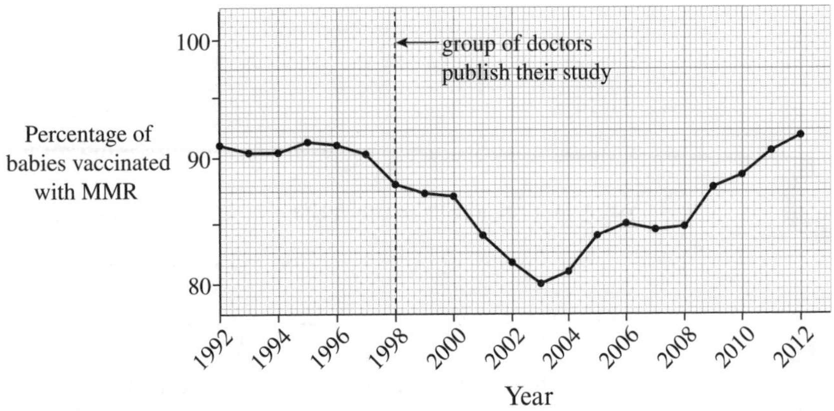

(a) Identify the year in which the rate of MMR vaccination was lowest.

.... **(1 mark)**

(b) Predict what would happen to the number of children suffering from measles in the period 1998–2004. Justify your answer.

....

.... **(2 marks)**

(c) The target immunisation rate for measles is 95%. Explain why it is not necessary for every child in the population to be immunised.

....

....

.... **(3 marks)**

Had a go ☐ Nearly there ☐ Nailed it! ☐

Antibiotics and painkillers

1 Penicillin is an example of an antibiotic.

(a) Explain why antibiotics can be used to treat bacterial infections in people.

....

.... (2 marks)

Guided

(b) A man has a very bad cold. Colds are caused by viruses. The man asks a pharmacist if he should take some penicillin to help cure his cold. Explain whether the pharmacist would advise the man to take penicillin.

The pharmacist's advice would be .... penicillin.

This is because antibiotics .... viruses. (2 marks)

2 Some hospitals have problems with a bacterium called MRSA. MRSA is resistant to many common antibiotics. The graph shows how many people died from MRSA infection between 2000 and 2008.

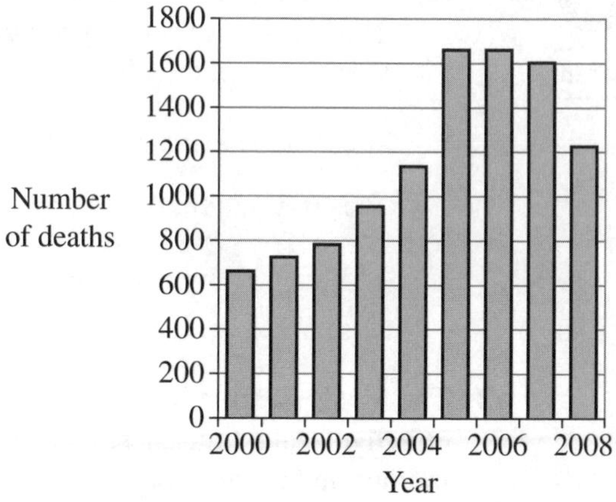

(a) Describe the pattern shown by the graph.

....

....

....

.... (2 marks)

(b) Suggest a reason to explain the change in the numbers of deaths after 2006.

....

.... (2 marks)

(c) Explain how a population of antibiotic-resistant bacteria, such as MRSA, might develop from non-resistant bacteria.

....

....

.... (3 marks)

3 Explain why it is difficult to develop drugs that kill viruses without also damaging the body's tissues.

Think about where viruses exist in the body.

....

.... (2 marks)

New medicines

1 Which of the following is a drug that was traditionally extracted from willow?
Tick **one** box.

aspirin ☐ digitalis ☐ insulin ☐ penicillin ☐ **(1 mark)**

2 The development of a new medical drug involves a series of tests. Testing can move to the next stage only if the substance has successfully passed the previous stage.

Guided

(a) Complete the table to show the correct order of stages in developing a new medical drug.

Stage	Order (1 to 5)
substance tested in a small number of healthy people	
discovery of a substance that may be a new medical drug	1
drug given widely by doctors to treat patients	
substance tested in cells and tissues in the lab	
substance tested in a large number of people with the disease it may treat	

(2 marks)

(b) Give **two** functions of a clinical trial in the development of a new medical drug.

....

.... **(2 marks)**

(c) Describe **two** stages of pre-clinical testing in the development of a new medical drug.

....

.... **(2 marks)**

3 Scientists trialled a new medical drug intended to lower blood pressure. Group A contained 1000 people with normal blood pressure and Group B contained 1000 people with high blood pressure. Half the volunteers in each group were given the new medicine and half were given a placebo. At the end of the trial, the scientists determined the number of volunteers in each group with high blood pressure.

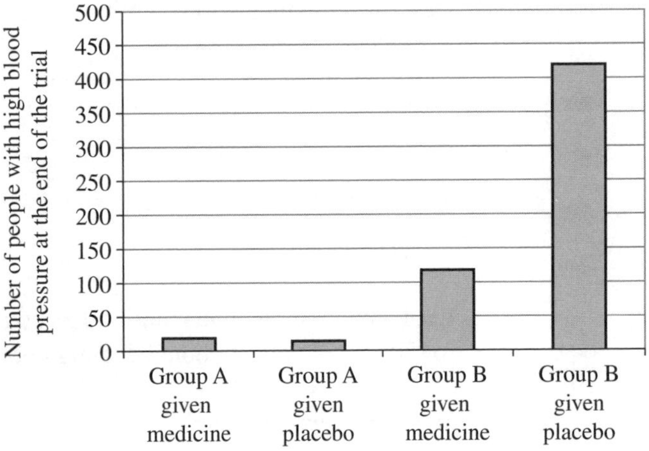

The graph shows the results.

(a) Describe what is meant by a placebo.

.... **(1 mark)**

(b) Use information from the graph to explain the effectiveness of this medical drug.

Think about whether the drug is effective and evidence for this.

....

.... **(2 marks)**

Had a go ☐ Nearly there ☐ Nailed it! ☐

Monoclonal antibodies

1 Monoclonal antibodies are produced from a single clone of cells.

(a) Give **two** advantages of cloning an antibody-producing cell.

1 ....

2 .... **(2 marks)**

Guided

(b) Describe how these antibody-producing cells are made before being cloned.

Include the names of each type of cell in your answer.

Cells called .... are collected from mice. The cells are stimulated to make a particular .... These cells are then fused with a .... cell to make a .... cell. **(4 marks)**

2 The diagram below shows a pregnancy test stick. It uses monoclonal antibodies to detect a hormone associated with pregnancy. The stick changes colour if high levels of this hormone are present.

region containing monoclonal antibodies to test for pregnancy hormone

handle

Give a reason why this stick would not detect a different hormone.

.... **(1 mark)**

3 Monoclonal antibodies can be used in the detection and treatment of cancer.

(a) Describe how monoclonal antibodies can be used to detect cancer cells in the body.

....

....

....

.... **(3 marks)**

(b) Explain how monoclonal antibodies could be used to kill cancer cells without harming healthy cells.

A toxic or radioactive substance could be involved.

....

....

....

.... **(4 marks)**

Plant disease

1 Bacterial wilt is a plant disease caused by a bacterium. Give the type of pathogen that causes the following:

(a) black spot

.... **(1 mark)**

(b) aphid infestation

.... **(1 mark)**

2 Plants can be damaged by a range of ion deficiency conditions. Complete the table below.

Guided

	Needed for	Symptom of deficiency
Magnesium ions		chlorosis (loss of green colour)
Nitrate ions		

(4 marks)

3 A farmer noticed that plants growing in one field had yellow leaves that curled at the edges and some of them had brown spots. He wondered if the plants were infected with tobacco mosaic virus.

(a) Give **one** symptom of plant disease, other than the ones seen by the farmer.

.... **(1 mark)**

(b) The farmer had not seen these symptoms before. Suggest a reason why he thought that the plants were infected with tobacco mosaic virus.

What sources could a gardener or farmer use to identify plant diseases?

.... **(1 mark)**

(c) Give a reason why the farmer cannot rely on visible symptoms alone to identify the disease.

.... **(1 mark)**

(d) The farmer noticed that most of the plants in the field were affected. The farmer consulted an agricultural adviser for help. Explain why the adviser asked the farmer to send her samples of:

(i) soil from several different places in the field

....

.... **(2 marks)**

(ii) affected plants

....

.... **(2 marks)**

Plant defences

Guided

1 Plants have physical defences to resist invasion by microorganisms. For example, tree bark and layers of dead cells around stems fall off, carrying microorganisms away.

Give **two** physical defences of leaves.

The cell membranes are surrounded by ....

The epidermal layer is covered by a .... **(2 marks)**

2 A student investigated whether garlic plants contain a substance that kills bacteria. She crushed a garlic clove in a little water. She added a little of this garlic–water mixture to a tube containing a bacterial culture. The student added water to a second tube containing a bacterial culture. She put lids on the tubes and examined the culture a few days later.

(a) State **two** factors that the student should keep constant in this investigation.

1 ....

2 .... **(2 marks)**

(b) Explain why the student repeated the investigation using water alone.

....

.... **(2 marks)**

(c) Give **one** advantage to the garlic plant of containing an antibacterial substance.

.... **(1 mark)**

3 The leaves of some Mimosa plants curl and droop when lightly touched.

normal

touched

(a) Suggest how this response is a mechanical adaptation to protect against insects.

....

.... **(2 marks)**

(b) Describe **one** other mechanical adaptation that helps to protect plants against animals.

> Remember that animals may try to eat the plants or lay their eggs on the plants.

.... **(1 mark)**

Extended response – Infection and response

Many infectious diseases can be treated using vaccines and antibiotics. Edward Jenner developed the first vaccine, against smallpox, in 1796. Alexander Fleming discovered the first antibiotic, penicillin, in 1928. Medicine has been transformed since then by the wide availability of these and other vaccines and antibiotics.

Compare the use of vaccines and antibiotics in the treatment of infectious diseases.

It may help if you make a brief plan before you start writing.

Remember that the command word **compare** requires you to describe the similarities and differences of both things given in a question, not just one of them. For example, you could compare:

- how vaccines and antibiotics work
- when they can be used
- what type of pathogen they are effective against
- their benefits and drawbacks.

(6 marks)

Had a go ☐ Nearly there ☐ Nailed it! ☐

Photosynthesis

1 What is photosynthesis? Tick **one** box.

an exothermic process in which energy is transferred to the chloroplasts by light ☐

an endothermic process in which energy is transferred to the chloroplasts by light ☐

an exothermic process in which energy is transferred from the chloroplasts by light ☐

an endothermic process in which energy is transferred from the chloroplasts by light ☐

(1 mark)

2 Photosynthesis can be represented by this equation: $6CO_2 + 6H_2O \rightarrow C_6H_{12}O_6 + 6O_2$

Use these chemical symbols to write the corresponding word equation.

.... **(1 mark)**

3 Plants have many uses for the carbohydrate produced by photosynthesis.

(a) Some of this carbohydrate is converted into substances for use in energy stores. Name **one** of these storage substances.

.... **(1 mark)**

(b) Some of this carbohydrate is converted into substances that are not used in energy stores. Name **one** of these products and describe what the plant uses it for.

This question is about a different use of the sugars from the use described in part (a), so make sure that you pick a product that is not used for energy storage!

Product: ....

Use: .... **(2 marks)**

4 The rate of photosynthesis may be affected by the carbon dioxide concentration and temperature.

(a) Give **two** other factors that may affect the rate of photosynthesis.

1 ....

2 .... **(2 marks)**

Guided

(b) Sketch a graph to show how the rate of photosynthesis may be affected by carbon dioxide concentration, and a graph to show how the rate of photosynthesis may be affected by temperature.

A sketch is an approximate drawing. You do not need values on your graphs but you should label the axes.

Rate of photosynthesis

Carbon dioxide concentration

(4 marks)

Limiting factors

1 Describe what is meant by a limiting factor.

Guided

This is a factor or variable that stops the rate of something ....

The rate will increase only if this factor is .... (2 marks)

2 The graph shows how the rate of photosynthesis changes with light intensity. The data show the rate at three different concentrations of carbon dioxide.

(a) Describe how changing the concentration of carbon dioxide affects the rate of photosynthesis.

....

.... (2 marks)

(b) Give **one** way in which the rate of photosynthesis could be increased further than the highest rate shown on the graph.

.... (1 mark)

(c) Tomato growers often increase the concentration of carbon dioxide in their greenhouses.

Explain how this will affect the yield of tomatoes.

....

....

.... (3 marks)

3 A farmer wants to increase the rate of growth of his strawberry plants. He changes the temperature of his greenhouse from 15 °C to 25 °C and notices that the plants grow more quickly. They grow at the same rate at 35 °C, but do not grow at all at 45 °C. The farmer knows that photosynthesis uses enzymes. Explain why the growth rate of the plants changes in this way.

Remember to include reasons when you explain something.

....

....

....

.... (4 marks)

Had a go ☐ Nearly there ☐ Nailed it! ☐

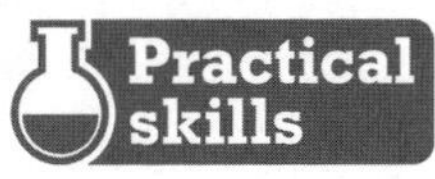

Investigating photosynthesis

1 A student investigated how the rate of photosynthesis in pondweed changed with light intensity. She placed a lamp at different distances from pondweed in a test tube. She counted the number of bubbles produced in 1 minute. The table shows her results.

Distance, *d*, in m	0.30	0.25	0.20	0.15	0.10
$\frac{1}{\text{distance}}$	3.3				10
$\frac{1}{(\text{distance})^2}$	11				100
Number of bubbles	14	32	54	72	88

(a) For each distance, complete the table to show:

Maths skills The distances are precise to 2 significant figures, so your final answers should be given to this too.

Guided

(i) 1/distance

(1 mark)

(ii) $1/(\text{distance})^2$

(1 mark)

(b) Plot a graph to show the number of bubbles on the *y*-axis and $1/(\text{distance})^2$ on the *x*-axis.

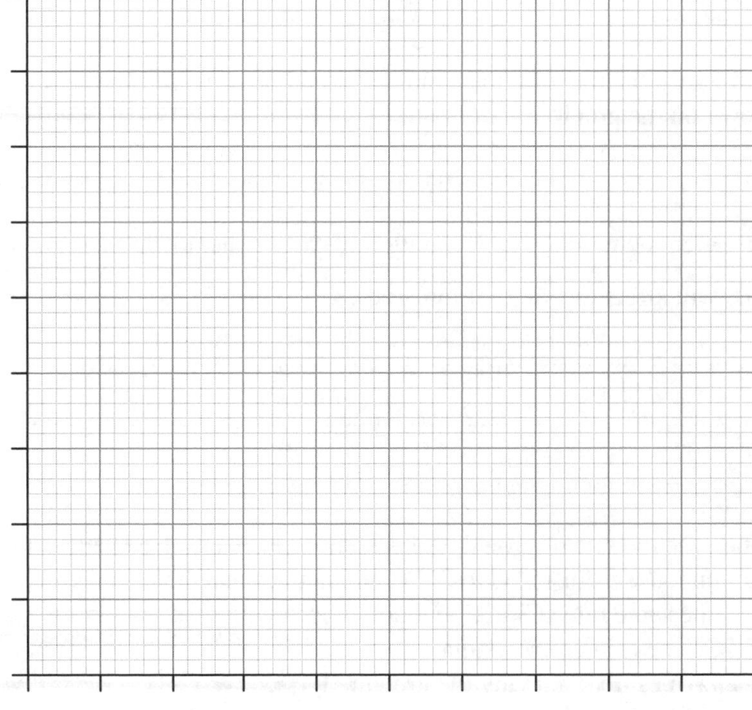

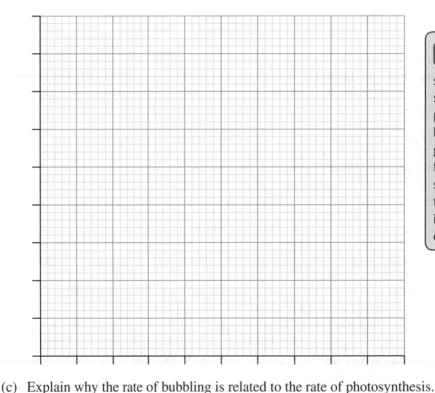

Maths skills Choose suitable scales for the axes. These should allow the plotted points to cover at least half the area of the graph. Draw a line of best fit. This can be curved or straight, depending on the data, but should ignore points that are clearly anomalies.

(4 marks)

(c) Explain why the rate of bubbling is related to the rate of photosynthesis.

....

....

.... **(3 marks)**

(d) The value for $1/(\text{distance})^2$ is proportional to the light intensity. Describe the relationship between light intensity and rate of photosynthesis.

....

.... **(2 marks)**

Respiration

1 Which of these is the word equation for aerobic respiration? Tick **one** box.

glucose → lactic acid ☐

glucose → ethanol + carbon dioxide ☐

glucose + oxygen → lactic acid + water ☐

glucose + oxygen → carbon dioxide + water ☐ **(1 mark)**

2 Energy is transferred by respiration in cells.

(a) Give **three** uses for this energy in animals.

Guided

keeping warm, ...

.... **(3 marks)**

(b) Explain whether respiration is an exothermic reaction or an endothermic reaction.

Say what type of reaction it is, and why.

....

.... **(2 marks)**

3 Plants use energy transferred by light for photosynthesis. Suggest an explanation for why plants need to respire continuously.

....

.... **(2 marks)**

4 Anaerobic respiration can take place in yeast cells.

(a) Give another name for this process.

.... **(1 mark)**

(b) Anaerobic respiration can also happen in muscle cells. Compare anaerobic respiration in a yeast cell with anaerobic respiration in a muscle cell.

Describe the similarities and differences between these two processes.

....

....

.... **(3 marks)**

(c) Aerobic respiration and anaerobic respiration can happen in muscle cells. Compare these two processes.

....

....

....

.... **(4 marks)**

Had a go ☐ Nearly there ☐ Nailed it! ☐

Responding to exercise

1 The graph shows the pulse rate of an athlete at rest, and after 5 minutes of different types of exercise.

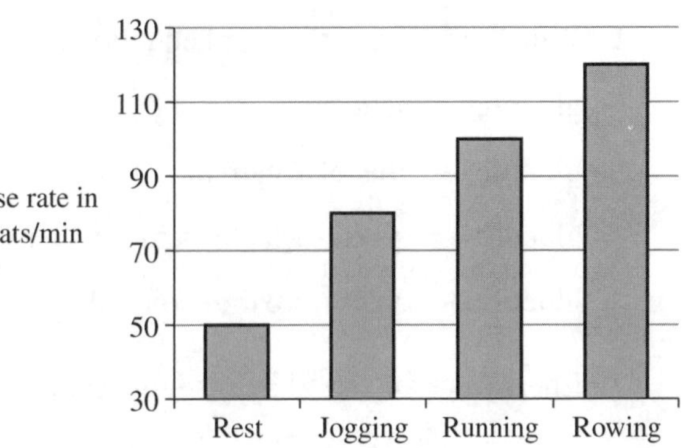

(a) Calculate the percentage increase in pulse rate between jogging and running.

100 – 80 = ... beats / min

percentage increase =

(.......... /80) × 100 = ... % **(2 marks)**

(b) Explain why the pulse rate increases as the amount of activity increases.

The pulse rate is the same as the heart rate.

....

....

.... **(3 marks)**

2 The graph shows how oxygen consumption changes before, during and after exercise. The intensity of the exercise kept increasing during the period marked 'Exercise'.

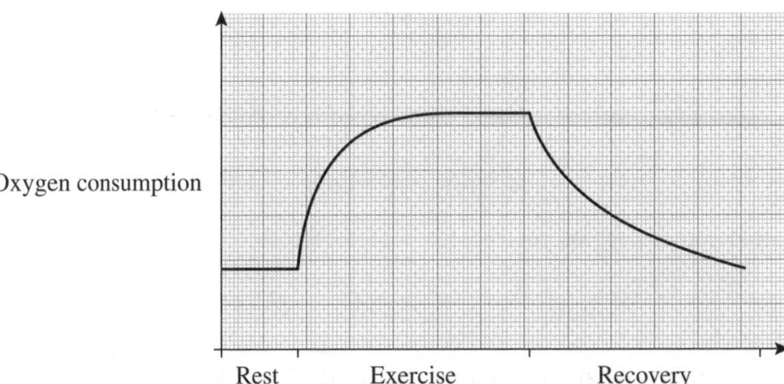

(a) Explain the shape of the graph during the period marked 'Exercise'.

....

....

.... **(3 marks)**

(b) Explain the shape of the graph during the period marked 'Recovery'.

....

....

.... **(3 marks)**

Had a go ☐ Nearly there ☐ Nailed it! ☐

Metabolism

1 What is metabolism? Tick **one** box.

the rate at which food is digested to soluble molecules ☐

how quickly respiration happens in mitochondria in cells ☐

how efficiently waste substances are removed from a cell or body ☐

the sum of all the reactions in a cell or the body ☐ **(1 mark)**

Guided

2 The liver is an important organ in the body. Its cells can carry out many different conversions of one substance to another. Give **three** examples of these conversions.

1 Lactic acid (produced during anaerobic respiration) into ....

2 Excess glucose into .... for storage in the liver.

3 Excess amino acids into .... (then into ....). **(3 marks)**

3 Glucose can be used to synthesise amino acids in living organisms.

(a) Name the ion needed to form amino acids using glucose.

Remember that glucose and other carbohydrates do not contain nitrogen atoms, but amino acids do.

.... **(1 mark)**

(b) Name the type of complex substance made from many amino acids.

.... **(1 mark)**

(c) Glucose molecules can be combined to form complex carbohydrates. Name **two** of these complex carbohydrates formed by plants.

One of these compounds is used for storage.

1 ....

2 .... **(2 marks)**

4 Lipids are fats and oils.

(a) Name the **two** different substances that react together to form lipids.

1 ....

2 .... **(2 marks)**

(b) Describe how many molecules of each substance named in part (a) are needed to form **one** lipid molecule.

.... **(1 mark)**

5 Name **three** types of digestive enzyme, and their substrates.

A **substrate** is the substance acted on by an enzyme.

1 ....

2 ....

3 .... **(3 marks)**

Had a go ☐ Nearly there ☐ Nailed it! ☐

Extended response – Bioenergetics

Photosynthesis and aerobic respiration are important processes in plants. They involve different reactants and products, and take place in different parts of the cell.

Compare photosynthesis and aerobic respiration. In your answer, you should include equations, descriptions of energy transfer and relevant sub-cellular structures.

> It may help if you make a brief plan before you start writing.
>
> Remember that the command word **compare** requires you to describe both things given in a question, not just one of them. So take care to describe:
>
> - the similarities between the two processes
> - the differences between the two processes.
>
> If you are not confident with balanced chemical equations, give word equations as part of your answer.
>
> Make sure that you do not include anaerobic respiration in your answer, because this is not required here.

....

....

....

....

....

....

....

....

....

....

....

....

....

....

....

....

....

....

....

.... **(6 marks)**

Homeostasis

1 Explain what is meant by homeostasis.

Guided

Homeostasis is the regulation of the .... of

a cell or organism to maintain ....

in response to .... changes. **(3 marks)**

2 Homeostasis uses automatic control systems, which may involve nervous responses or chemical responses. These systems include three main parts.

(a) Describe the function of receptor cells.

....

.... **(2 marks)**

(b) The pancreas can act as a coordination centre. Name **one** other coordination centre.

.... **(1 mark)**

(c) Homeostatic systems include effectors.

(i) Name **two** types of effectors.

1 ....

2 .... **(2 marks)**

(ii) Describe what effectors do in a homeostatic control system.

....

.... **(2 marks)**

3 Human body temperature is controlled so it remains close to 37 °C.

(a) Give **two** other body conditions that are controlled.

1 ....

2 .... **(2 marks)**

(b) Give a reason that explains why body temperature should not be allowed to decrease very much.

Think about the effect of temperature on chemical reactions.

....

....

.... **(2 marks)**

(c) Give a reason that explains why body temperature should not be allowed to increase very much.

Think about the effect of temperature on enzyme activity.

....

....

.... **(2 marks)**

Had a go ☐ Nearly there ☐ Nailed it! ☐

Neurones

1 The nervous system contains three main types of neurones. Sensory neurones carry electrical impulses from receptors to the central nervous system or CNS (the brain and spinal cord). Describe the function of:

(a) relay neurones

.... **(1 mark)**

(b) motor neurones

.... **(1 mark)**

2 The diagram shows a sensory neurone connected to receptor cells in the skin.

Guided

Explain how the sensory neurone is adapted to its function.

What do the axon terminals, myelin sheath, dendrites, and axon and dendron do?

The axon and dendron are long so the neurone can ....

....

....

....

.... **(4 marks)**

3 The table shows the speeds at which nerve impulses are transmitted through different kinds of neurones.

Myelin sheath	Impulse speed in m/s
present	25
absent	3

(a) Explain the effect of the presence of the myelin sheath.

Remember to include what the effect is, using information from the table.

....

.... **(2 marks)**

(b) Multiple sclerosis (MS) is a disorder in which the myelin sheath surrounding neurones in the spinal cord is destroyed. Explain the effect this would have on the movement of a person with MS.

....

.... **(2 marks)**

Reflex actions

1 The diagram shows a junction where neurone **X** meets neurone **Y**.

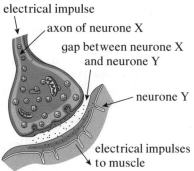

electrical impulse
axon of neurone X
gap between neurone X and neurone Y
neurone Y
electrical impulses to muscle

(a) Give the name of the junction between two neurones.

.... **(1 mark)**

(b) Identify which neurone, neurone **X** or neurone **Y**, is a motor neurone. Give a reason for your answer.

.... **(1 mark)**

(c) Describe how signals pass from neurone **X** to neurone **Y**.

Guided

Neurone X releases a ....

which..,

causing .... **(3 marks)**

2 Choose **three** words from the box to complete the sentence about the features of reflex actions.

innate	slow	automatic	conscious	learned	rapid

Reflex actions are ...., ..., and .... **(3 marks)**

3 The diagram shows components of a reflex arc.

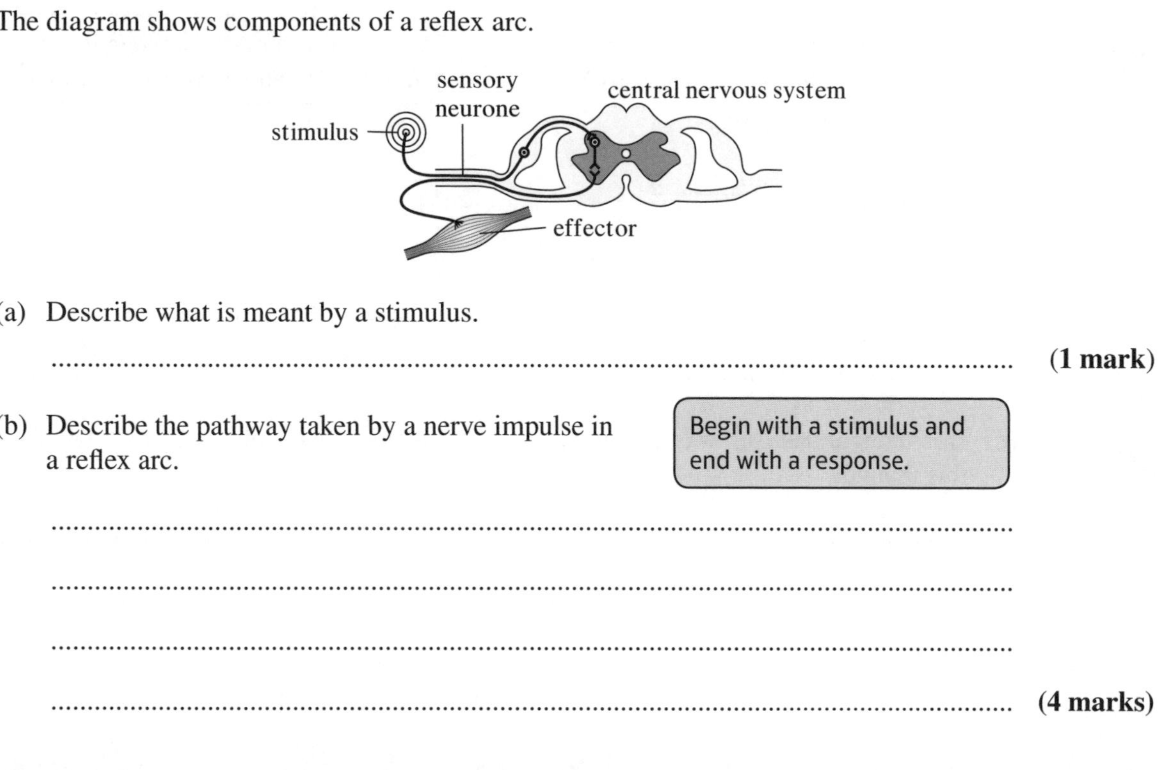

(a) Describe what is meant by a stimulus.

.... **(1 mark)**

(b) Describe the pathway taken by a nerve impulse in a reflex arc.

Begin with a stimulus and end with a response.

....

....

....

.... **(4 marks)**

Had a go ☐ Nearly there ☐ Nailed it! ☐

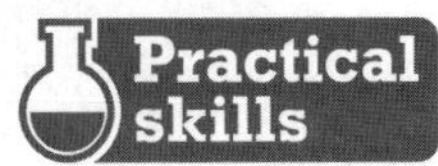

Investigating reaction times

1 Two students carried out an investigation into human reaction times. This is the method they used.

> Ann sat with her forearm resting on the bench, with her hand over the edge. Bob held a 30-cm ruler so that Ann's thumb was just touching the ruler, level with the zero mark at the bottom. Bob then dropped the ruler and Ann caught it. The distance the ruler dropped before she managed to catch it was recorded.

The students repeated the experiment several times. The table shows their results.

Drop number	1	2	3	4	5	6
Drop distance in mm	193	186	190	184	181	176

(a) Calculate the mean distance.

Maths skills Add the values together then divide the total by the number of values.

....

....

.... mm **(1 mark)**

(b) The reaction time can be determined from the drop distance. One way to do this involves the equation:

$$\text{reaction time in s} = \sqrt{\frac{\text{drop distance in cm}}{491}}$$

(i) Identify the result in the table that corresponds to Ann's quickest reaction time. Give a reason that explains your answer.

....

.... **(2 marks)**

Guided

(ii) Calculate Ann's reaction time for drop 1. Give your answer to 2 significant figures.

Remember to convert from mm to cm, and to use the equation given to you.

distance = $\frac{\dots\dots}{10}$ = ... cm

....

....

.... s **(3 marks)**

(c) Bob thought that Ann was getting better with practice. Give a reason that explains this thought.

....

.... **(2 marks)**

The brain

1 The diagram shows a cross-section through the human brain. Three regions are labelled **X**, **Y** and **Z**.

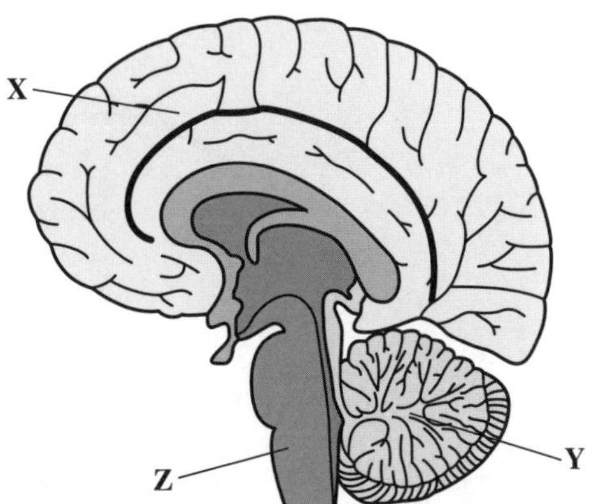

(a) Which row in the table correctly identifies all three regions of the brain? Tick **one** box.

The word 'cortex' comes from the Latin word for 'bark' or outer layer.

Cerebral cortex	Cerebellum	Medulla	
Y	X	Z	☐
X	Y	Z	☐
X	Z	Y	☐
Z	X	Y	☐

(1 mark)

(b) Describe **one** function of each of the following parts of the brain:

(i) medulla

.... **(1 mark)**

(ii) cerebellum

.... **(1 mark)**

(c) Describe **two** functions of the cerebral cortex.

1 ....

2 .... **(2 marks)**

2 Neuroscientists have been able to map the regions of the brain to particular functions.

Guided

(a) Describe **three** ways in which neuroscientists can do this.

The scientists can electrically .... different parts of the brain. They can also study people with .... and use .... **(3 marks)**

(b) Explain why it is very difficult to investigate brain disorders.

....

.... **(2 marks)**

The eye

1 Label the diagram of the eye to identify the parts labelled **A**, **B**, **C**, **D** and **E**.

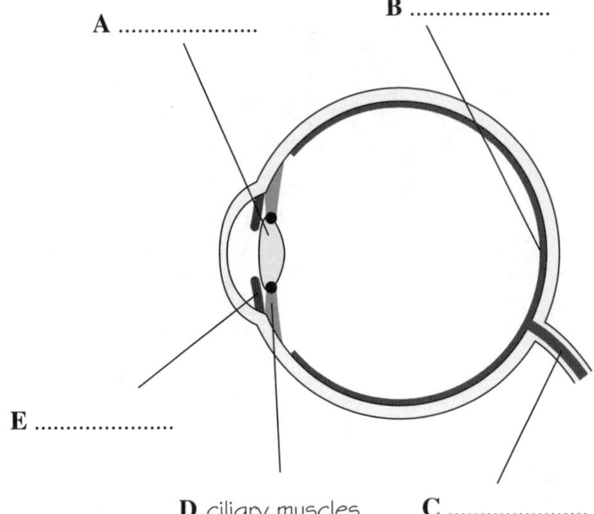

(5 marks)

2 Describe how the structure of the retina is related to its function.

Think about the structure of the retina, and what the retina does.

....

....

.... **(3 marks)**

3 Light rays entering the eye are focused to form a sharp image on the retina.

(a) Describe the path taken by a light ray through the eye.

....

....

.... **(3 marks)**

(b) Describe the changes that happen in the eye as it changes focus from a distant to a near object.

....

....

.... **(3 marks)**

4 An optometrist may use eyedrops to temporarily dilate the pupils in a patient's eyes. This lets the optometrist examine the back of the eyes better. Explain how the patient's vision may be affected until the effects of the drops wear off.

Think about what 'dilates' means.

....

....

.... **(3 marks)**

Eye defects

1 Short-sightedness (myopia) and long-sightedness (hyperopia) are two common eye defects.

Complete the table by placing a tick (✓) in a box against each statement to show whether it applies to a person with myopia or to a person with hyperopia.

Statement	Myopia	Hyperopia
Near objects appear clear but distant objects appear blurred.		
The eyeball is shorter than it should be.		
The lens cannot become curved enough to correct the problem.		

(3 marks)

2 An optometrist prescribes spectacles for a patient. The diagram shows the type of lens in these spectacles.

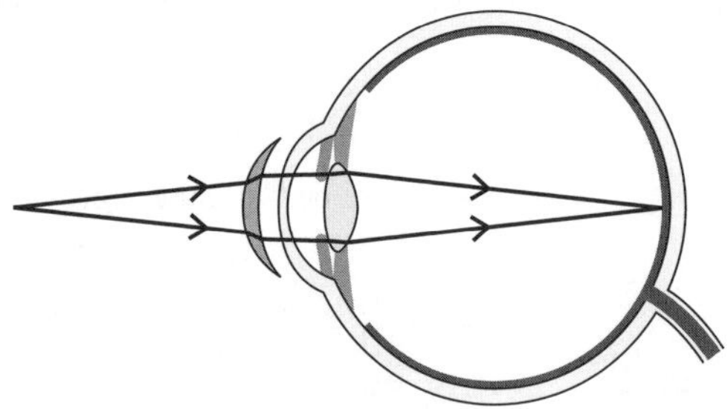

(a) Give **two** other ways in which defects in vision can be treated.

1 ....

2 .... **(2 marks)**

(b) Name the eye defect corrected by the lens, and give a reason for your answer.

Look at the shape of the lens or how it refracts light passing through it.

.... **(1 mark)**

(c) Explain how the lens improves the patient's vision.

Guided

The lens refracts the light before it enters the eye, so that it ....

....

.... **(2 marks)**

(d) A patient with a different eye defect is prescribed different spectacles. Draw lines to show how the light rays continue and become focused.

Your diagram should be like the one above, but the rays will refract differently as they pass through the spectacle lens.

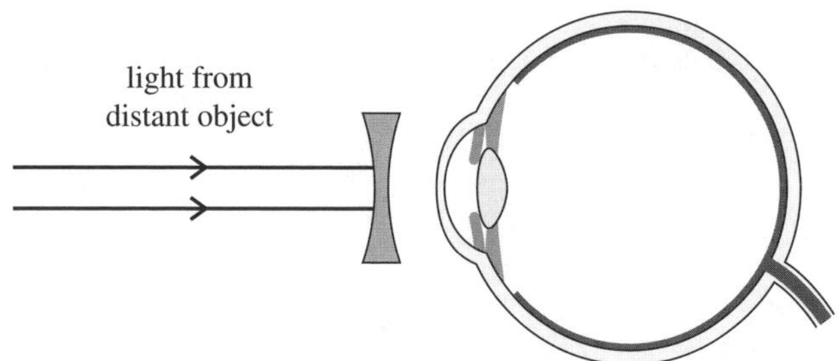

(3 marks)

Had a go ☐ Nearly there ☐ Nailed it! ☐

Thermoregulation

1 The average temperature during a January day in the UK is around 3 °C. Your body temperature will start to drop if you go outside without warm clothing.

(a) Describe how the brain detects a decrease in skin temperature.

....

.... **(2 marks)**

(b) You may start to shiver as your body temperature falls. Explain how shivering helps to warm you up.

....

.... **(2 marks)**

2 An investigation was carried out into the effect of exercise on core (internal) body temperature and the temperature of the skin's surface. The graph shows the results of this investigation.

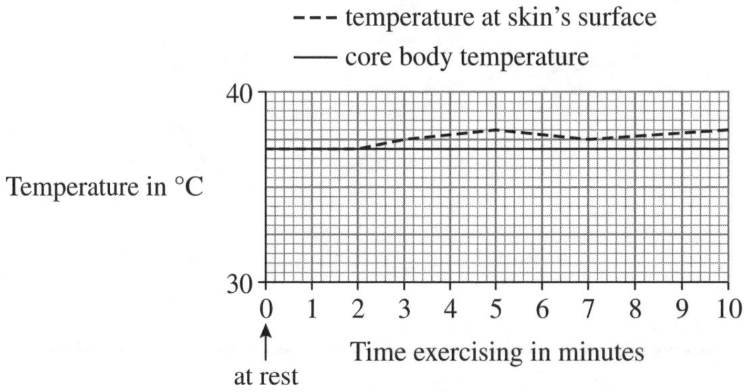

--- temperature at skin's surface

—— core body temperature

Temperature in °C

40

30

0 1 2 3 4 5 6 7 8 9 10

Time exercising in minutes

at rest

(a) Compare what happens to the skin and body temperature in the investigation.

You should describe the similarities and / or differences between the temperatures in the two places.

....

.... **(2 marks)**

(b) Explain how the sweat glands help to cool the body.

....

.... **(2 marks)**

Guided

(c) Explain how blood vessels in the skin cause a rise in the temperature of the skin's surface during exercise.

When the temperature rises, blood vessels ....

There is more ....

This means that more ....

.... **(3 marks)**

Hormones

1 Label the diagram to identify the endocrine glands **A** to **F**.

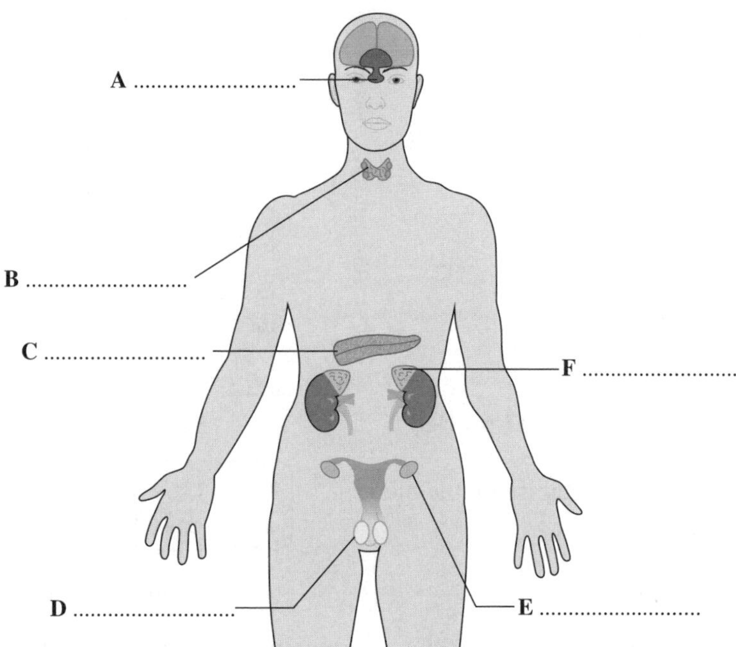

(6 marks)

2 Complete the table to show where each hormone is produced, with **one** example of its target organ.

Guided

Hormone	Produced in	Target organ
ADH		
adrenaline	adrenal gland	heart / muscles
FSH and LH		
glucagon		
oestrogen		
progesterone		
testosterone		
TSH		

(8 marks)

3 Compare the endocrine system with the nervous system.

Think about how signals travel, and the speed and duration of their effects.

....

....

....

....

.... **(3 marks)**

Had a go ☐ Nearly there ☐ Nailed it! ☐

Blood glucose regulation

1 The graph is an example of changes in blood glucose concentration during the day.

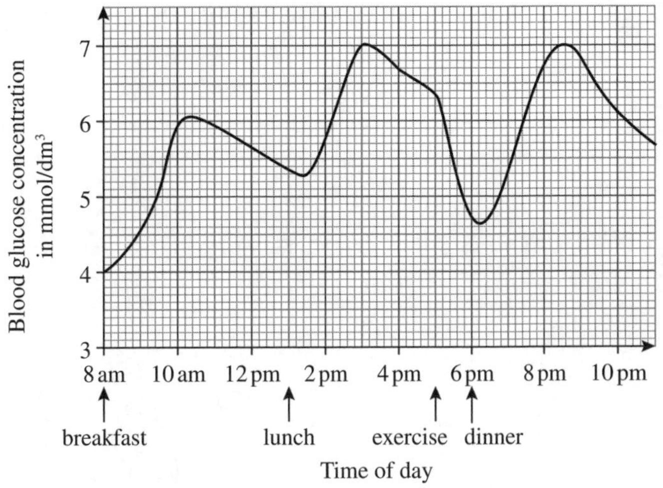

(a) Describe the trend in blood glucose concentration just after each meal is eaten.

....

....

....

....

(2 marks)

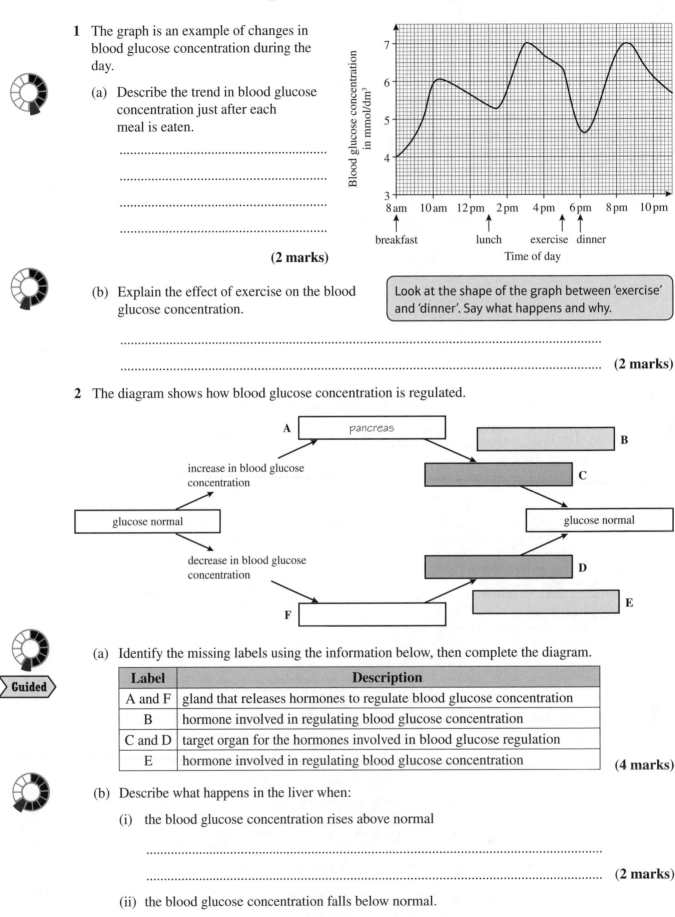

(b) Explain the effect of exercise on the blood glucose concentration.

Look at the shape of the graph between 'exercise' and 'dinner'. Say what happens and why.

....

.... **(2 marks)**

2 The diagram shows how blood glucose concentration is regulated.

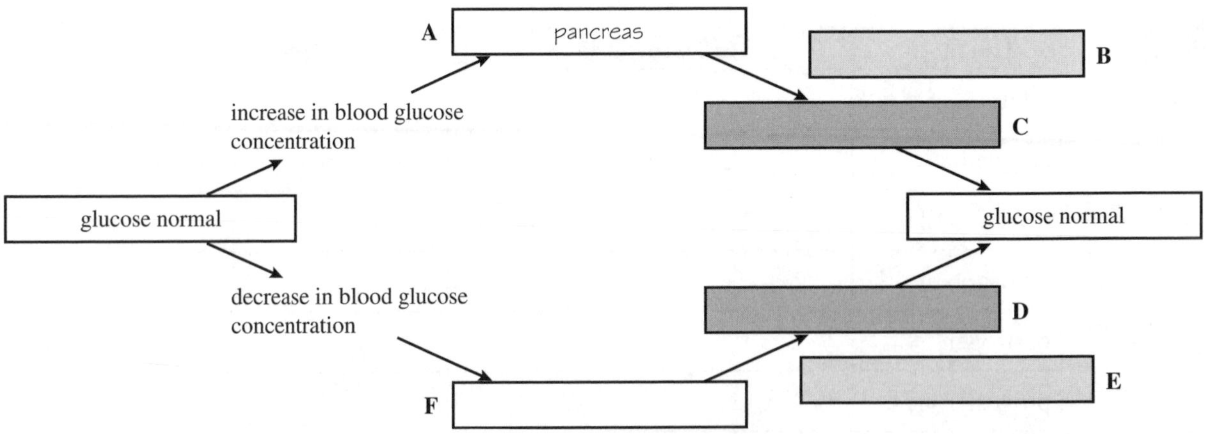

(a) Identify the missing labels using the information below, then complete the diagram.

Guided

Label	Description
A and F	gland that releases hormones to regulate blood glucose concentration
B	hormone involved in regulating blood glucose concentration
C and D	target organ for the hormones involved in blood glucose regulation
E	hormone involved in regulating blood glucose concentration

(4 marks)

(b) Describe what happens in the liver when:

(i) the blood glucose concentration rises above normal

....

.... **(2 marks)**

(ii) the blood glucose concentration falls below normal.

....

.... **(2 marks)**

Diabetes

1 A population of 1 864 035 people was sampled to estimate the occurrence of type 2 diabetes. The results were divided into groups according to body mass index (BMI). People who have a BMI greater than 30 are obese. The bar chart shows the results.

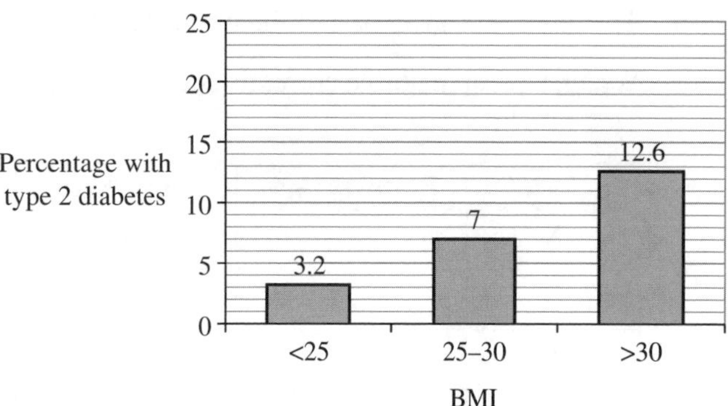

Guided

(a) Use the information to calculate the number of people in this population with type 2 diabetes.

Total percentage with type 2 diabetes = 3.2 + 7 + 12.6 = ... %

Number of people = ... × 1864 035/100

= .... **(2 marks)**

(b) Suggest a reason why the number calculated in part (a) may not be accurate.

Do not suggest an error in your calculation as a reason.

.... **(1 mark)**

(c) Determine what the results in the bar chart show.

Look for a trend and use the values shown to inform your answer.

....

.... **(2 marks)**

2 Diabetes exists as type 1 and type 2.

(a) Compare the causes of type 1 diabetes and type 2 diabetes.

....

.... **(2 marks)**

(b) Explain how type 1 diabetes is treated.

....

.... **(2 marks)**

(c) Describe **two** ways in which type 2 diabetes may be controlled.

....

.... **(2 marks)**

Controlling water balance

1 The diagram shows part of the urinary system.

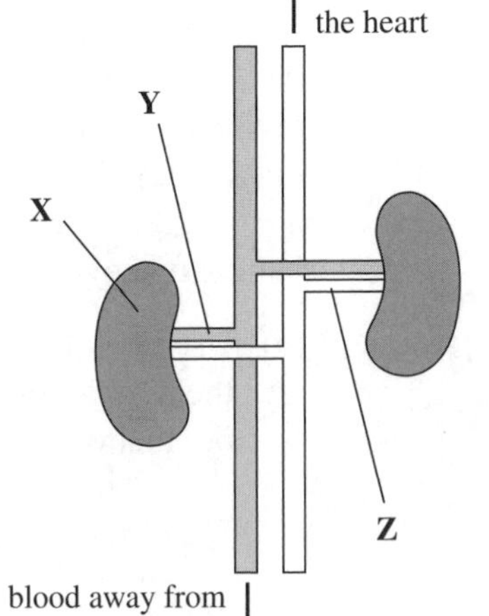

Which row in the table correctly identifies the three labelled parts? Tick **one** box.

Kidney	Renal artery	Renal vein	
X	Y	Z	☐
X	Z	Y	☐
Y	Z	X	☐
Z	Y	X	☐

(1 mark)

2 Give **two** other ways, other than in urine, in which water leaves the body.

1 ....

2 .... **(2 marks)**

3 Predict what would happen to red blood cells washed in water, and give a reason for your answer.

Think about how and why water moves across partially permeable membranes.

....

.... **(2 marks)**

4 The table shows the typical concentrations of some substances in blood plasma and urine.

Substance	Concentration in g/dm^3	
	Blood plasma	Urine
salts	9	20
protein	80	0
glucose	1	0
urea	0.3	18

(a) Explain why the concentration of water and salts in urine may differ over time.

....

....

.... **(2 marks)**

Guided

(b) Explain the differences in the concentration of:

(i) protein: Plasma protein molecules are too .... to pass through the .... **(2 marks)**

(ii) glucose: ....

.... **(2 marks)**

(iii) urea: ....

.... **(2 marks)**

ADH and the kidneys

1 Excess amino acids in the body need to be excreted safely.

(a) Give a reason why excess amino acids may be produced in the body.

.... **(1 mark)**

(b) Excess amino acids are deaminated in the liver.

(i) Name the nitrogen-containing substance formed in this process.

.... **(1 mark)**

(ii) Explain why this substance is then converted to urea.

....

.... **(2 marks)**

2 The water level in the body is controlled by the hormone ADH.

(a) Where in the body is ADH produced? Tick **one** box.

kidneys ☐ pituitary gland ☐

adrenal gland ☐ thyroid gland ☐ **(1 mark)**

(b) ADH acts on certain structures in the kidneys.

(i) Name these structures.

.... **(1 mark)**

(ii) Explain the effect of decreasing ADH levels on the volume of urine.

Give the effect of **less** ADH on the structures and why this changes the volume of urine.

....

....

.... **(3 marks)**

3 A student took part in a 5-mile run on a hot day. He did not take any water with him to drink.

Guided

(a) At the end of the run, the student needed to urinate. Describe how his urine would differ from that of a student who had drunk water during the same run.

The volume of urine would be ....

and the urine would be more .... **(2 marks)**

(b) Explain why the level of ADH changed in the student's blood during the run.

....

....

....

.... **(4 marks)**

Had a go ☐ Nearly there ☐ Nailed it! ☐

Kidney treatments

1 Kidney failure may be treated by dialysis with a kidney dialysis machine or by a kidney transplant. The chart shows the employment status of people who have had different forms of treatment for kidney failure.

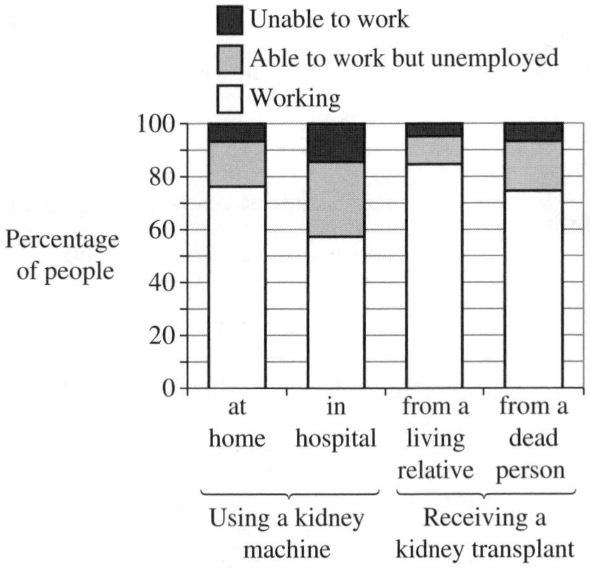

(a) Identify the form of treatment that allows the largest percentage of people to be able to work.

....

.... **(1 mark)**

(b) Give the percentage of people who use a kidney machine in hospital and are unable to work.

> **Maths skills** Work out the percentage represented by the 'unable to work' section.

.... **(1 mark)**

(c) Look at the data for people receiving treatment with a kidney machine. Suggest an explanation for why the place where a kidney machine is used may affect a patient's employment status.

....

....

.... **(3 marks)**

2 Describe the function of the partially permeable membrane in a kidney dialysis machine.

....

.... **(2 marks)**

Guided

3 The table shows the concentrations of some substances before and after kidney dialysis.

Complete the table by placing a tick (✓) against each correct comparison.

	Less than	The same as	More than	
Glucose concentration in dialysis fluid before is		✓		the concentration in blood after dialysis.
Glucose concentration in dialysis fluid after is				the concentration in blood after dialysis.
Urea concentration in dialysis fluid before is				the concentration in blood before dialysis.
Urea concentration in blood before is				the concentration in blood after dialysis.

(4 marks)

Reproductive hormones

1 Reproductive hormones cause secondary sex characteristics to develop during puberty. Oestrogen is the main female reproductive hormone.

(a) Name the main male reproductive hormone.

.... **(1 mark)**

(b) Give **one** effect the hormone named in part (a) has on the testes.

.... **(1 mark)**

(c) Give **two** effects the main reproductive hormones have on both boys and girls at puberty.

1 ....

2 .... **(2 marks)**

2 The diagram below shows typical timings of some features in the menstrual cycle.

(a) Identify the process happening between days 1 and 5, labelled X in the diagram.

.... **(1 mark)**

Guided

(b) Give the name and function of:

(i) LH

Name: ....

Function: stimulates the release of an egg. .... **(2 marks)**

(ii) FSH

Name: ....

Function: .... **(2 marks)**

(c) Identify the process happening around day 14, labelled Y in the diagram.

.... **(1 mark)**

(d) Name the **two** hormones involved in maintaining the lining of the uterus.

You will need to name a hormone not given in the diagram.

....

.... **(2 marks)**

Had a go ☐ Nearly there ☐ Nailed it! ☐

Control of the menstrual cycle

1 Four different hormones interact in the control of the menstrual cycle.

Guided

Complete the table by placing a tick (✓) in a box against each hormone to show where it is produced.

Hormone	Site of production		
	Ovaries	Pituitary gland	Empty egg follicle
oestrogen	✓		
FSH			
LH			
progesterone			

(4 marks)

2 The diagram shows how hormone levels change during the menstrual cycle.

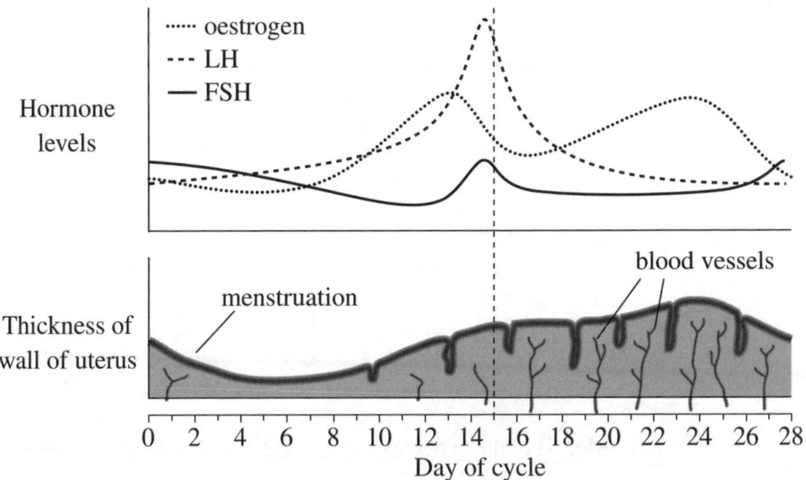

(a) FSH stimulates the growth and maturation of egg follicles. The production of a hormone is stimulated by this process. Identify this hormone.

Name one of the four hormones shown in the table in question 1.

.... **(1 mark)**

(b) Suggest, using information in the diagram, the effect of increasing levels of oestrogen on:

Say whether release is stimulated or inhibited, and give a reason why.

(i) the release of LH

.... **(1 mark)**

(ii) the release of FSH

.... **(1 mark)**

(c) The level of progesterone (not shown in the diagram) rises after ovulation. Suggest the effect of this on the release of FSH and LH after ovulation, and give a reason for your answer.

.... **(1 mark)**

(d) Explain whether the diagram uses data from a woman who became pregnant at day 16.

....

.... **(2 marks)**

Contraception

1 Fertility can be controlled by a variety of hormonal and non-hormonal methods of contraception.

(a) Name the surgical method used to sterilise men.

.... **(1 mark)**

(b) Barrier methods of contraception prevent the sperm reaching an egg.

(i) Name **two** barrier methods of contraception.

1 ....

2 .... **(2 marks)**

(ii) Suggest a reason why spermicidal agents may improve the effectiveness of these methods.

.... **(1 mark)**

2 The table shows some information about two oral contraceptives. Both inhibit release of FSH and LH.

	Mini pill	Combined pill
Contents	progesterone	oestrogen, progesterone
Success rate	96–99%	>99%

Guided

(a) The success rates assume that the pills are taken as prescribed. Explain why a woman is more likely to become pregnant if she does not take her pills regularly.

FSH causes an egg to .... and LH causes an egg to .... A missed pill causes the level of progesterone to ...., so ....

.... **(4 marks)**

(b) Progesterone may also be delivered by using implants. Suggest an advantage of this method of contraception over the mini-pill, and give a reason for your answer.

....

.... **(2 marks)**

(c) Similar to any medical drug, oral contraceptives may have side effects. The table shows the results of a study into blood clotting in 100 000 women.

	Non-smokers, not on 'the pill'	Non-smokers, pregnant	Smokers, not on 'the pill'	Non-smokers, on 'the pill'
Number of women with blood clotting	8	85	100	40

Evaluate the risk to women who are considering using this oral contraceptive.

Use the information to consider evidence for and against using this oral contraceptive.

....

....

.... **(3 marks)**

Had a go ☐ Nearly there ☐ Nailed it! ☐

Treating infertility

1 FSH and LH are hormones involved in the menstrual cycle.

(a) Describe the effect of LH.

Guided

LH stimulates .... **(1 mark)**

(b) There is an increase in body temperature at the same time that the level of LH increases. Explain how a woman could use this information to increase her chance of conceiving a child.

....

.... **(2 marks)**

(c) FSH and LH may be given to a woman in a 'fertility drug'. She may then become pregnant in the normal way. Explain a risk of this fertility treatment.

Think about the effect of FSH and LH on eggs in the ovary.

....

.... **(2 marks)**

2 *In-vitro* fertilisation (IVF) treatment can be used to help couples who are unable to conceive a child, for example, because of blocked oviducts in the woman or if the man produces few healthy sperm cells.

(a) Give a reason why the woman is given injections of FSH at the start of IVF treatment.

.... **(1 mark)**

(b) Describe the steps taken in IVF treatment after injections of FSH and LH.

....

....

....

.... **(4 marks)**

3 In 2010, 45 250 women underwent IVF treatment in the UK. Of these women, 12 400 were successful in having a child. The cost of one cycle of IVF treatment was £2500.

Use this information, and your own knowledge, to describe the benefits and drawbacks of IVF treatment.

....

....

....

.... **(4 marks)**

Adrenaline and thyroxine

1 Thyroxine plays an important role in growth and development.

(a) Name the gland that produces thyroxine.

.... **(1 mark)**

(b) Give the effect of thyroxine on the basal metabolic rate.

.... **(1 mark)**

2 Thyroxine production is controlled by TSH produced in the pituitary gland.

(a) Give the effect of low thyroxine levels on the levels of TSH.

.... **(1 mark)**

(b) Explain how control of thyroxine concentration in the blood is an example of negative feedback.

....

.... **(2 marks)**

3 A man is walking through a forest at dusk and hears a wolf howl.

(a) Explain why the concentration of adrenaline in the man's bloodstream increases.

> Include the source of adrenaline and a general situation in which adrenaline may be released.

....

.... **(2 marks)**

(b) The increased adrenaline affects the man's body.

(i) Give the effect of adrenaline on the heart rate.

.... **(1 mark)**

(ii) Describe the consequences of the effect given in part (i).

Guided

The delivery of .... and .... to the brain and .... increases. This prepares the man's body for .... **(3 marks)**

(c) Explain why the control of adrenaline levels is not an example of negative feedback.

....

....

.... **(3 marks)**

Plant hormones

1 The diagram shows how a plant shoot responds to light.

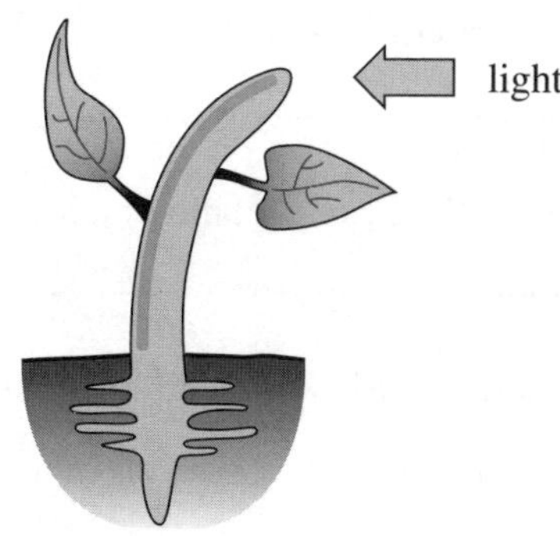

(a) What is this response called? Tick **one** box.

positive phototropism ☐

negative gravitropism ☐

negative phototropism ☐

positive gravitropism ☐. **(1 mark)**

(b) Label the diagram with an **X** to show a part of the plant where auxins are produced. **(1 mark)**

2 Compare the effects of high concentrations of auxins on cells in shoots and roots.

Guided

High concentrations of auxins cause elongation of cells in ....

but inhibit .... **(2 marks)**

3 The diagram shows a root emerging from a germinating seed.

(a) Give the direction in which the root will grow. Give a reason for your answer.

Include the stimulus to which the root responds in your answer.

....

.... **(2 marks)**

(b) Give **two** reasons why the response described in part (a) is important for roots.

1 ....

2 .... **(2 marks)**

(c) Explain how auxins affect the growth of the root.

....

....

.... **(3 marks)**

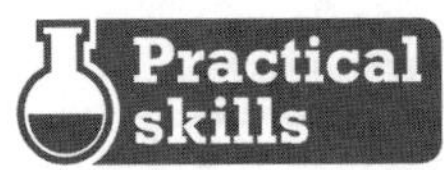

Investigating plant responses

1 Gibberellins are plant hormones. A group of students investigated the effect of gibberellins on the growth of pea plants. They used two sets of three plants, each 60 cm tall. The students sprayed one set with gibberellins dissolved in water and another set with water alone. The table shows the students' results.

	Height of pea plants in cm	
	Sprayed with gibberellins	Sprayed with water
plant 1	102	65
plant 2	93	61
plant 3	99	63

(a) The mean height of the plants sprayed with water was 63 cm.

Calculate the mean height of the plants sprayed with gibberellins.

Maths skills Add the values together then divide the total by the number of values.

....

....

Mean height = ... cm **(1 mark)**

(b) Describe the effect of gibberellins on the pea plants.

Make sure that you use the data in your answer.

....

.... **(2 marks)**

(c) Explain why one set of plants was sprayed with water.

....

.... **(2 marks)**

2 A teacher showed an experiment to her class. She took two growing shoots and removed the tip from one of them. She placed the shoots in a box, with a strong light source on the right-hand side. The diagrams show the experiment when the teacher had set it up, and after a few days.

light source

shoot with the tip cut off

a few days

Explain the results of this experiment.

Auxin is made in the tip of the shoot, so the shoot with its tip removed ....

....

....

.... **(4 marks)**

Had a go ☐ Nearly there ☐ Nailed it! ☐

Uses of plant hormones

1 People use plant hormones to control plant growth and development.

(a) Which plant hormone is used to control ripening of fruit? Tick **one** box.

auxin ☐

ethene ☐

gibberellin ☐

(1 mark)

(b) Describe **two** advantages to the food industry of controlling fruit ripening during storage and transport.

....

.... **(2 marks)**

2 Selective weedkillers contain auxins.

(a) Predict the effect of spraying plants with a solution containing one of these selective weedkillers. Give reasons that explain your answer.

Consider the effects on roots, as well as on shoots.

....

....

.... **(3 marks)**

(b) Crops such as wheat and barley have narrow leaves, but many weed plants have broad leaves. Broad-leaved plants absorb more auxins than narrow-leaved plants. Explain how this helps to increase crop yields when selective weedkillers are applied.

....

....

.... **(3 marks)**

3 Many plants, such as hydrangeas, normally flower in July and August. However, growers at the annual Chelsea Flower Show in May want to display hydrangea plants in full flower. Describe how a grower could produce hydrangea plants that flower in time for this show.

....

.... **(2 marks)**

4 Explain why plant growers may dip cuttings from a single plant in rooting powder.

Guided

Rooting powder contains ...., which are plant hormones that cause cuttings to develop .... This means that the growers can

.... **(3 marks)**

Extended response – Homeostasis and response

A patient was concerned about his weight loss and visited his doctor. The patient told his doctor that his diet and levels of exercise had not changed very much. The doctor diagnosed hyperthyroidism, an overactive thyroid gland.

Explain why an overactive thyroid gland could cause the patient to have a decreased body mass.

> It may help if you make a brief plan before you start writing.
>
> Think about what the thyroid does, including the hormone it produces and how levels of this hormone are normally controlled. Make sure that you clearly link the overactive thyroid gland to the loss of weight, and use correct terminology in your answer.

(6 marks)

Had a go ☐ Nearly there ☐ Nailed it! ☐

Meiosis

1 Describe the process of fertilisation in humans.

Guided

A male gamete (a .... cell) .... with ....

.... to form a .... **(2 marks)**

2 The diagram shows a cell with two pairs of chromosomes undergoing meiosis.

parent cell

A

Describe what is happening during the stage labelled **A**.

.... **(1 mark)**

3 The diagram shows a cell about to divide by meiosis.

In the space below, draw the daughter cells that are produced from this cell.

Your diagram should show the correct number of cells, with the correct number of chromosomes in each one.

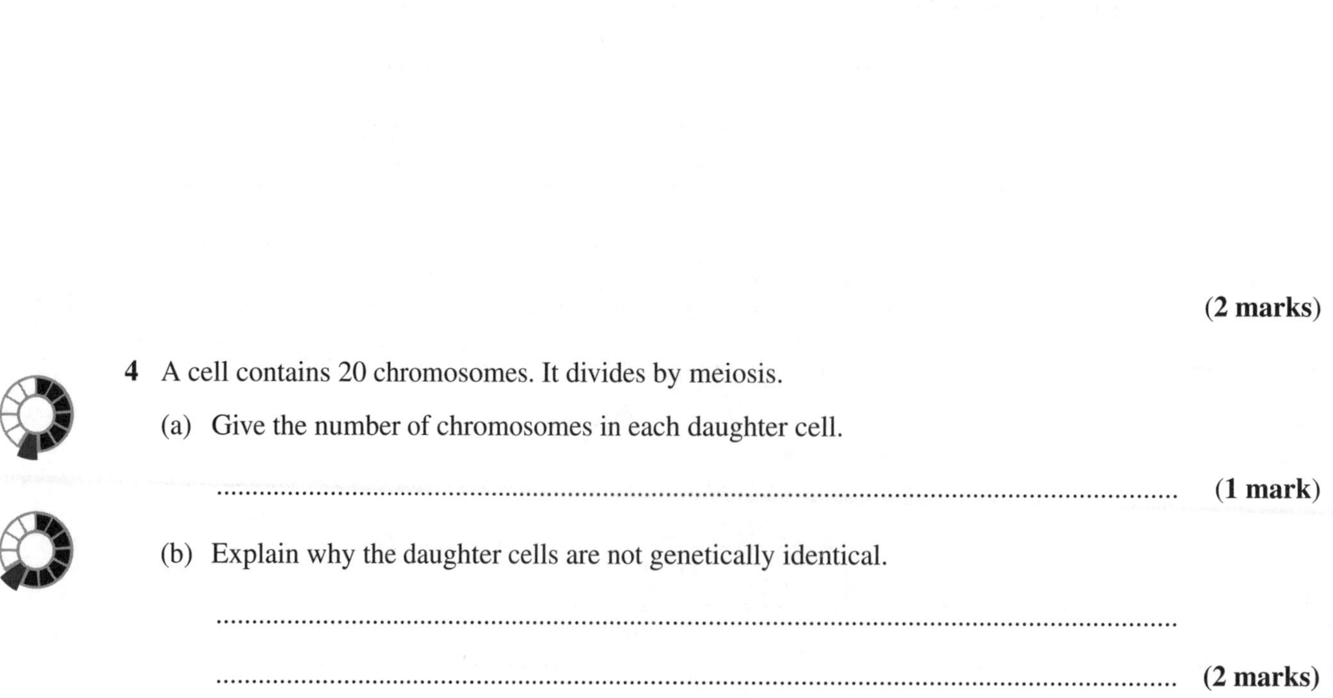

(2 marks)

4 A cell contains 20 chromosomes. It divides by meiosis.

(a) Give the number of chromosomes in each daughter cell.

.... **(1 mark)**

(b) Explain why the daughter cells are not genetically identical.

....

.... **(2 marks)**

Sexual and asexual reproduction

1 Complete the table to compare features of sexual and asexual forms of reproduction.

Guided

	Sexual reproduction	Asexual reproduction
Need to find a mate		
Mixing of genetic information	mixes genetic information from each parent	no mixing of genetic information
Characteristics of offspring		

(3 marks)

2 Strawberry plants produce runners during the growing season. A new plant develops where a runner touches the ground. Later in the summer the original plant produces flowers. These are fertilised and produce fruits with seeds. Animals eat the fruits and leave them in their faeces, far from the original plant.

(a) Explain which of these is sexual reproduction and which is asexual reproduction.

Make sure you identify each type of reproduction and give a reason for your answer.

Runners ....

....

Fruits ....

.... **(4 marks)**

(b) Give **one** benefit and **one** drawback to the strawberry plant of:

(i) asexual reproduction

....

....

.... **(2 marks)**

(ii) sexual reproduction

....

....

.... **(2 marks)**

3 Describe **two** ways in which sexual reproduction requires organisms to expend more energy than asexual reproduction.

....

....

.... **(2 marks)**

Had a go ☐ Nearly there ☐ Nailed it! ☐

DNA and the genome

1 The water level in the body is controlled by the hormone ADH. An ADH molecule is made of 9 amino acids. How many bases are in the length of DNA that codes for ADH? Tick **one** box.

9 ☐ 18 ☐ 27 ☐ 36 ☐ **(1 mark)**

2 Chromosomes contain genetic information in the form of DNA.

(a) Give the word used to describe all the DNA of an organism.

.... **(1 mark)**

(b) Describe, in terms of DNA, what is meant by a gene.

Guided

A gene is a small section of DNA which ....

.... **(2 marks)**

3 DNA is a polymer.

(a) Describe the shape of the structure formed by the two strands in a DNA molecule.

.... **(1 mark)**

(b) The diagram shows a small section of DNA.

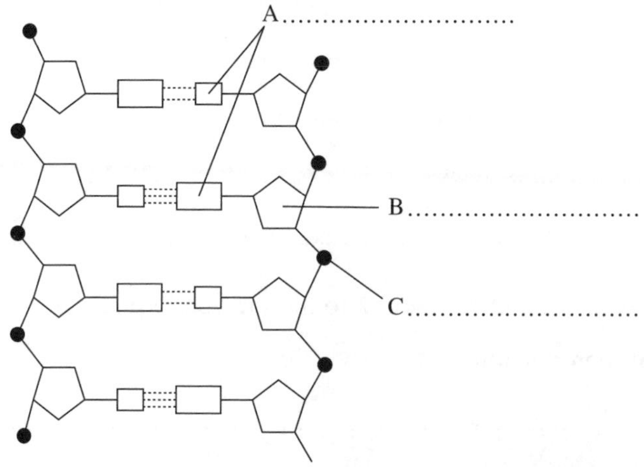

A..............................

B..............................

C..............................

(i) Identify the components A, B and C of the DNA structure. Write your answers on the diagram.

You do not need to name any bases in your answers.

(3 marks)

(ii) Give **one** piece of evidence from the diagram that DNA is a polymer.

.... **(1 mark)**

4 The sequence of bases on part of one strand of DNA was ATGGGC.

Give the order of the bases on the complementary strand of DNA, and explain your answer.

Order of bases ....

Explanation ....

.... **(3 marks)**

Protein synthesis

Guided

1 The synthesis of a protein involves amino acids attached to specific carrier molecules. Synthesis happens in several stages.

(a) Complete the table to show the correct order of stages in protein synthesis, starting with 1.

Stage	Order (1 to 6)
Template molecule binds to a ribosome.	
Template molecule leaves the nucleus.	
Ribosome moves along the template.	
Carrier molecule with an amino acid attaches to the template.	3
Carrier molecule is released.	
Amino acid joins to the growing protein chain.	

(3 marks)

(b) After it leaves a ribosome, a completed protein chain changes so that it can carry out its function. Describe this change.

....

.... **(2 marks)**

2 Mutation affects DNA.

Guided

(a) Give **three** ways in which DNA may be changed by mutation.

1 There can be a small change in the sequence of bases in the DNA.

2 ....

3 .... **(3 marks)**

(b) Explain how a mutation in non-coding DNA can change the phenotype of an organism.

Non-coding parts of DNA do not code for proteins. An individual's phenotype is its observed characteristics or traits.

....

.... **(2 marks)**

3 Cystic fibrosis is caused by a mutation that produces an inactive protein in the lungs. Explain how a mutation leads to production of an inactive protein.

....

....

....

.... **(3 marks)**

Had a go ☐ Nearly there ☐ Nailed it! ☐

Genetic terms

1 Describe what is meant by:

(a) a gamete

Include what it is and the process by which it forms.

....

.... (2 marks)

(b) alleles

....

.... (2 marks)

2 Explain the difference between the terms **genotype** and **phenotype**.

Guided

The genotype of an organism is the alleles of a ....

However, an organism's phenotype is its ....

produced by .... (3 marks)

3 People with brown or blue eyes have different combinations of two alleles. The recessive allele b codes for blue eyes, and the dominant allele B codes for brown eyes.

(a) Some people are heterozygous for eye colour.

(i) Give the heterozygous genotype for eye colour.

Use the symbols b and/or B in your answer.

.... (1 mark)

(ii) Explain the eye colour of a person with the heterozygous genotype.

Give the eye colour and a reason for your answer.

....

.... (2 marks)

(b) A girl has blue eyes. Explain what her genotype must be.

....

.... (2 marks)

4 Pea plants form smooth or wrinkled seeds. RR and Rr plants form smooth seeds and rr plants form wrinkled seeds. Explain which allele is dominant and which is recessive.

....

....

....

.... (4 marks)

Genetic crosses

1 *Drosophila* are fruit flies. Most *Drosophila* have normal wings, but some can have poorly formed wings. The allele for normal wings is D, and the allele for poorly formed wings is d. The d allele is recessive.

Drosophila A *Drosophila* B

(a) The diagram shows two different *Drosophila*. Identify the alleles present in *Drosophila* B, and give a reason for your answer.

....

.... **(2 marks)**

(b) Two *Drosophila* with the alleles Dd are mated together to produce offspring.

Guided

Draw a Punnett square diagram to determine the probability of an offspring having badly formed wings. Identify any offspring with poorly formed wings.

In your completed diagram, circle the genotypes of the offspring with badly formed wings.

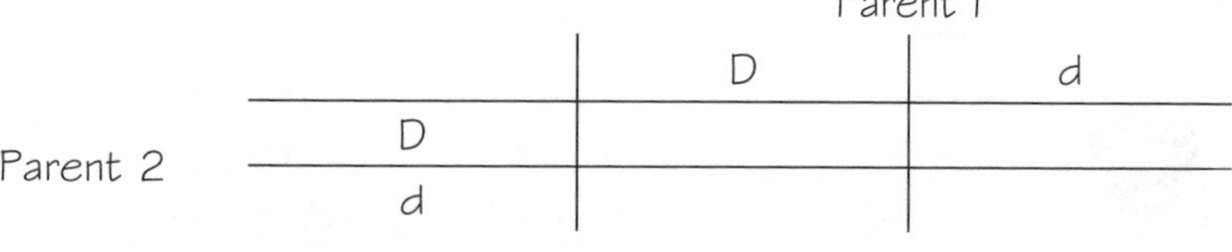

Probability .... **(4 marks)**

2 Fur colour in mice can be represented by two alleles, G and g. Two parent mice produced a total of 40 offspring. Half of these offspring were white, which is the recessive characteristic, and the rest were grey.

(a) Draw a Punnett square diagram to show this cross.

(3 marks)

(b) Give the genotypes of the mice with grey fur and of the mice with white fur.

Grey fur ....

White fur .... **(1 mark)**

Had a go ☐ Nearly there ☐ Nailed it! ☐

Family trees

1 Sickle-cell anaemia is a condition caused by a recessive allele. Two healthy parents have a child who has sickle-cell anaemia. Which one of the following statements is correct? Tick **one** box.

Both parents are homozygous for the sickle-cell allele. ☐

One parent is homozygous for the sickle-cell allele and the other is homozygous for the normal allele. ☐

Both parents are heterozygous for the sickle-cell allele. ☐

One parent is heterozygous for the sickle-cell allele and the other is homozygous for the normal allele. ☐

(1 mark)

2 Cystic fibrosis (CF) is a genetic disorder that affects cell membranes. It is caused by a recessive allele. The diagram shows a family tree for a family with some affected individuals.

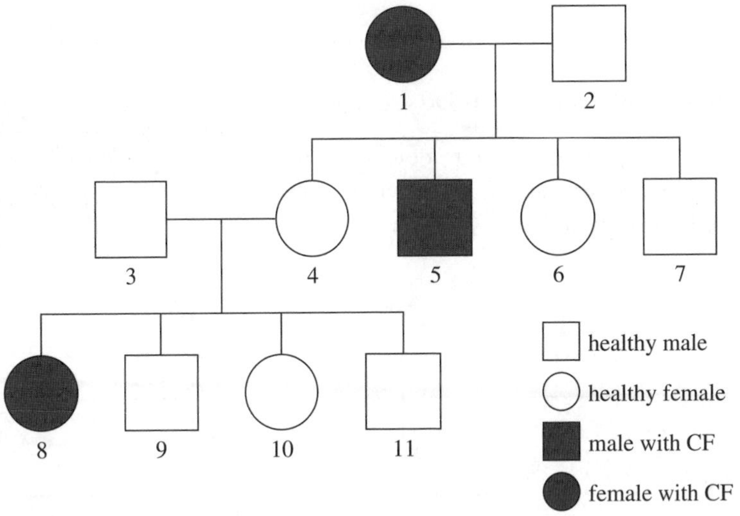

1 2

3 4 5 6 7

8 9 10 11

healthy male

healthy female

male with CF

female with CF

(a) Give the number of cystic fibrosis alleles that must be inherited for an individual to have the disorder.

.... **(1 mark)**

(b) Determine the number of males in the family who have a homozygous recessive genotype.

.... **(1 mark)**

(c) Determine the genotype of person 4. Explain your answer using F for the normal allele and f for the recessive allele.

Guided

Person 4 doesn't have cystic fibrosis but they must have inherited an .... allele from their mother. So they must have inherited an .... allele from their father. This means that person 4's genotype is .... **(3 marks)**

(d) Explain how the family tree provides evidence that cystic fibrosis is caused by a recessive allele.

Look for patterns of inheritance in the family tree to work out how CF is inherited.

....

.... **(2 marks)**

Inheritance

1 (a) Give the number of pairs of chromosomes found in the nucleus of a human body cell.

.... **(1 mark)**

(b) Complete the Punnett square diagram to show how sex is determined in humans.

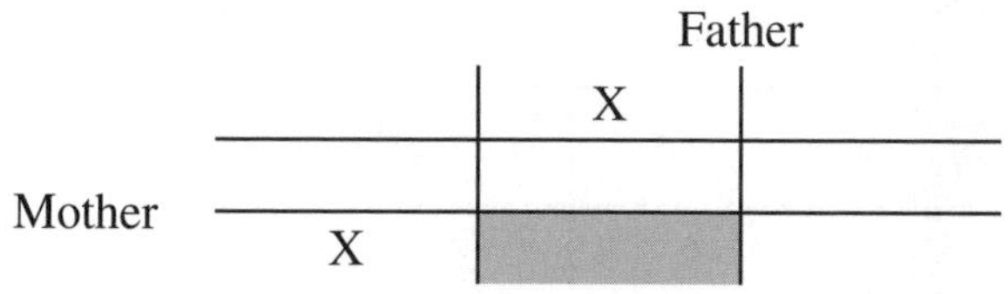

		Father	
		X	
Mother			
	X	(shaded)	

(2 marks)

(c) Give the sex of a child represented by the genotype in the shaded box.

.... **(1 mark)**

Guided

2 Embryo screening may be used to test for the presence of certain inherited disorders. This allows parents to make informed decisions about an embryo or pregnancy. The graph shows how the chance of having a child with Down's syndrome varies with the age of the mother.

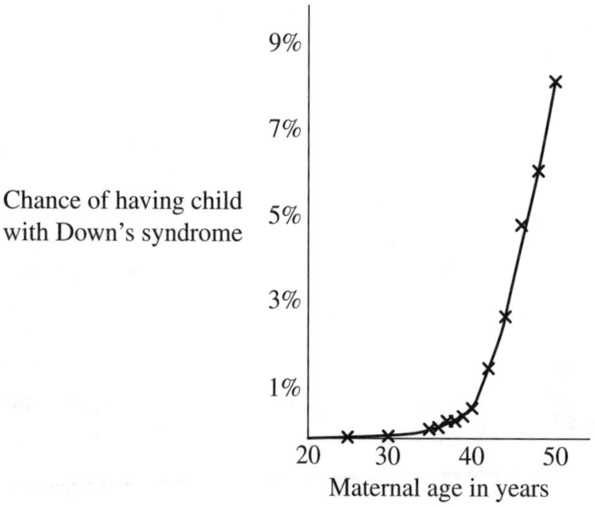

Use the information in the graph to suggest why women of different ages are given different advice on embryo screening for Down's syndrome.

Chance of having child with Down's syndrome: 1%, 3%, 5%, 7%, 9%

Maternal age in years: 20, 30, 40, 50

The graph shows that the chance of a woman having a child with Down's syndrome .... as the woman's age .... So pregnant women over the age of 40 are .... likely to be offered embryo screening. **(2 marks)**

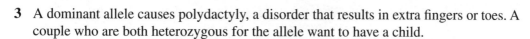

3 A dominant allele causes polydactyly, a disorder that results in extra fingers or toes. A couple who are both heterozygous for the allele want to have a child.

Draw a Punnett square diagram to determine the probability of the child **not** having polydactyly.

Use the symbol P for the dominant allele and p for the recessive allele.

Identify any children without polydactyly.

Circle the appropriate genotype in your diagram or write it down.

Probability = ... **(4 marks)**

Had a go ☐ Nearly there ☐ Nailed it! ☐

Variation and evolution

Guided

1 All variants in a population arise from mutations. Complete the table to show the influence of mutations on phenotypes. Place a tick in each correct box.

	Mutations		
	Very few	Some	Most
Have no effect on phenotype			
Influence the phenotype		✓	
Determine the phenotype			

(3 marks)

2 Variation may have genetic causes, environmental causes, or a combination of genetic and environmental causes. Give the type of cause of the following:

(a) a scar on the skin

.... (1 mark)

(b) differences between a pair of identical twins

.... (1 mark)

(c) body mass in a class of Year 11 students

.... (1 mark)

3 Explain why, when an environment changes, some individuals within a population survive and reproduce, whereas others do not.

....

.... (2 marks)

4 Suggest a reason to explain why forming a new species by natural selection is a slow process.

....

.... (2 marks)

5 The giraffe is the tallest living land animal. Adult giraffes are 4.5–5 m tall, and nearly half of this height is due to their necks. These long necks allow giraffes to reach leaves for food in high trees. Giraffes are believed to have evolved from short-necked creatures similar to antelopes.

Describe how the long neck of giraffes may have evolved by natural selection over many generations.

Remember that there is variation in a population and differences in the characteristics of individuals.

....

....

....

.... (4 marks)

Selective breeding

1 Describe what is meant by selective breeding.

Guided

It is the process by which humans breed ....

for particular .... **(2 marks)**

2 Dairy farming involves producing milk from cows. Suggest a desirable characteristic that dairy farmers may wish to select for in their herd.

.... **(1 mark)**

3 Food production can be increased by selective breeding of crop plants.

(a) Suggest **two** different characteristics that could be selected for in a crop suitable for use in any country.

1 ....

2 .... **(2 marks)**

(b) Suggest **one** other characteristic that might be selected for in a crop to be grown in a country with a hot, dry climate.

.... **(1 mark)**

4 Wheat plants produce edible seeds at the end of thin stems. Artificial fertilisers used by farmers encourage wheat stems to grow very quickly. In the last century, long-stemmed wheat plants were crossed with short-stemmed dwarf varieties in a selective breeding programme.

(a) Suggest an advantage to wheat plants of having fast-growing long stems.

Think about how plants obtain the energy needed for photosynthesis.

.... **(1 mark)**

(b) Suggest a reason why plant breeders select wheat plants with short stems.

Think about the problems caused by heavy seed heads.

.... **(1 mark)**

5 Describe how pig breeders could use selective breeding to produce lean pigs (less body fat).

....

....

....

.... **(4 marks)**

6 Selective breeding can lead to 'inbreeding'. Give **two** drawbacks of this happening.

Inbreeding involves breeding between close relatives, which leads to a reduction in variation in a population.

....

.... **(2 marks)**

Had a go ☐ Nearly there ☐ Nailed it! ☐

Genetic engineering

1 Some potato plants have been genetically engineered so that they can resist attack by insect pests. Their cells contain a gene from a different organism that produces a toxic protein. What has genetic engineering done to these potato plants? Tick **one** box.

made no changes to their genome or phenotype ☐

changed their phenotype but not their genome ☐

changed their genome but not their phenotype ☐

changed their genome and their phenotype ☐

(1 mark)

Guided

2 Scientists have produced genetically modified mice that glow green in blue light. These 'glow mice' contain a gene naturally found in jellyfish. Describe how this genetically modified organism is produced.

The gene from a .... is cut out using ....

This gene is transferred to a .... embryo cell, and inserted into

a chromosome. The embryo is then allowed to develop as normal. **(3 marks)**

3 A student used the internet to research genetically modified crops. This is what she found out about rice.

> 'Golden rice' is a type of genetically modified rice. It is called golden because the gene for making beta-carotene has been inserted, which turns the rice yellow. Beta-carotene is needed for humans to produce vitamin A. Vitamin A deficiency causes a poor immune response and difficulty seeing at night.

(a) Explain why golden rice might be useful for countries where people have a poor diet.

....

.... **(2 marks)**

(b) Explain **one** reason why some people are opposed to the production of GM crops, such as golden rice.

There are different reasons why people may not agree with the growing of GM crops. Choose one and say why people might have this concern.

....

.... **(2 marks)**

4 People with Type 1 diabetes cannot produce insulin and need to inject themselves with this hormone. Until recently, insulin extracted from the pancreas of pigs was used. More recently, human insulin produced from GM bacteria has been used. Explain the advantages of using GM bacteria to produce the insulin for treating people with Type 1 diabetes.

....

....

....

.... **(4 marks)**

Stages in genetic engineering

1 Plant cells may be genetically engineered. One method involves using a modified virus to transfer a gene into plant cells. What is the function of the virus in this example of genetic engineering? Tick **one** box.

The correct answer to question 1 cannot be 'plasmid', because plasmids are found in bacteria.

pathogen ☐ host ☐

vector ☐ plasmid ☐ **(1 mark)**

2 Bacteria can be genetically engineered to produce human insulin. The flow chart shows some of the stages in this process.

Stage 1: Restriction enzyme used on human DNA	→	Stage 2: Restriction enzyme used to open plasmid from a bacterium	→	Stage 3: DNA ligase used to form the modified plasmid	→	Stage 4: Modified plasmid placed into bacteria and bacteria placed in a fermenter

(a) Two enzymes are used in this process: a 'restriction enzyme' and an enzyme called 'DNA ligase'. Identify which of these two enzymes is used as:

(i) a cutting enzyme

.... **(1 mark)**

(ii) a sticking or joining enzyme

.... **(1 mark)**

(b) Describe what is meant by a 'modified plasmid' (stage 3).

Remember to say what a plasmid is.

....

.... **(2 marks)**

(c) Explain why the same enzyme is used in stages 1 and 2.

Think about the feature of the cut ends of DNA that allow them to be 'sticky' so that they join together.

....

....

.... **(2 marks)**

Guided

(d) A fermenter is a container in which bacteria can grow and divide, producing very large numbers of bacteria. Suggest why it is sensible to have only bacteria with modified plasmids in the fermenter.

If there were non-GM bacteria in the fermenter as well, they might use up

.... or they might ....

.... **(2 marks)**

Cloning

1 A student researched the process of embryo transplanting, which is used to clone animals. This is what he found out.

> Sperm are used to fertilise egg cells. A fertilised egg is allowed to develop for a short time. Its cells are then separated before they become specialised, and the new individual embryos are transplanted into host mothers. The embryos then develop normally.

(a) Give a reason that explains why the embryo cells are separated before they become specialised.

....

.... **(2 marks)**

(b) Explain why the new individual embryos are genetically identical to each other but not to their host mothers.

....

.... **(2 marks)**

2 Tissue culture is used to produce clones of a plant.

(a) Describe what clones are.

.... **(1 mark)**

Guided

(b) Give **two** reasons why plants may be cloned.

to help to preserve rare plant species, and to ....

.... **(2 marks)**

(c) Describe how plant tissue culture is used to produce clones from a parent plant.

....

....

.... **(3 marks)**

3 A dog called 'Snuppy' was cloned from an adult dog by scientists in South Korea.

Describe the steps that the scientists would have taken when producing Snuppy.

> You need to outline the main stages in adult cell cloning.

....

....

....

....

....

.... **(5 marks)**

Darwin and Lamarck

1 In the early nineteenth century, most peppered moths were light coloured. This acted as camouflage when they rested on tree trunks, reducing the chance of being eaten by birds. Some peppered moths were dark and were less well camouflaged. Air pollution gradually turned the tree trunks darker. Scientists counted the numbers of light moths and dark moths over a 40-year period. The graph shows their results.

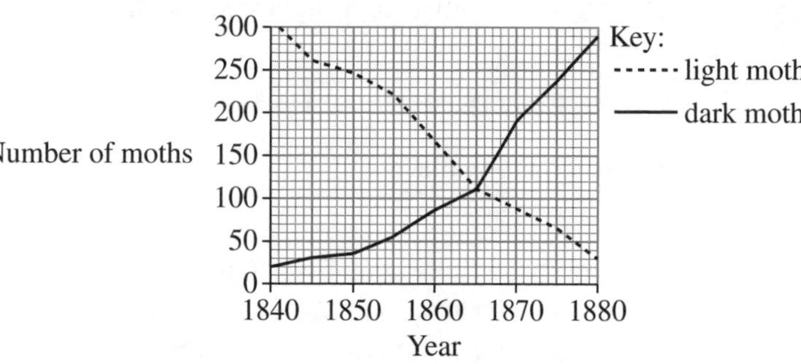

(a) Identify the year in which there was an equal number of light and dark moths.

.... **(1 mark)**

(b) Describe what the graph shows about the changes in the numbers of moth types.

....

.... **(2 marks)**

(c) About 10% of moths were light coloured in 1880. Suggest **two** reasons to explain why this form of the peppered moth did not disappear completely.

1 Not all trees were darkened, so some light moths were still camouflaged.

2 .... **(2 marks)**

2 Charles Darwin published his ideas about evolution in 1859. His ideas caused a lot of controversy.

(a) Describe **two** reasons why his theory of evolution by natural selection was accepted only gradually.

....

.... **(2 marks)**

(b) Give **two** ways in which Darwin was able to provide evidence in support of his theory.

....

.... **(2 marks)**

3 Apes have long arms that let them swing efficiently through the trees. Jean-Baptiste Lamarck proposed a different theory of evolution before the theory proposed by Darwin. Describe the differences between how Lamarck might have explained the evolution of long arms in apes, and how Darwin might have explained it.

> You do not need to give detailed explanations of both theories. Focus on the key differences.

....

....

....

.... **(4 marks)**

Had a go ☐ Nearly there ☐ Nailed it! ☐

Speciation

1 Which two scientists first suggested the theory of evolution by natural selection? Tick **one** box.

Lamarck and Darwin ☐ Darwin and Mendel ☐

Lamarck and Wallace ☐ Darwin and Wallace ☐ **(1 mark)**

2 The porkfish is a fish that lives in the seas between North America and South America. Millions of years ago, land started to emerge in this sea. A narrow strip of land (an isthmus) formed between the two continents about 3.5 million years ago. It is narrowest in modern-day Panama.

Guided

(a) The modern porkfish has developed through speciation into two different species of fish. One of these lives in the Caribbean Sea and the other in the Pacific Ocean. Explain what is meant by speciation.

Speciation is the process in which a population changes to form varieties

that cannot .... **(2 marks)**

(b) Explain why the formation of the isthmus led to changes in the population of porkfish.

....

....

.... **(3 marks)**

3 Salamanders are amphibians that look like lizards. Different salamander species of the genus *Ensatina* are found around the Central Valley of California. Each species of salamander can breed with its neighbouring species to produce infertile offspring, with one exception: in Cuyamaca State Park, *E. eschscholtzi* and *E. klauberi* cannot interbreed at all.

Suggest an explanation for why these different salamander species evolved.

Think about why the species change, going around the Central Valley, and then why the two species in the Cuyamaca State Park cannot interbreed.

....

....

....

.... **(4 marks)**

Mendel

1 Gregor Mendel (1822–1884) was the first person to discover the basics of inheritance. Here are some statements about his work. For each statement, explain how this aspect of Mendel's work allowed him to draw valid conclusions from his data.

(a) Mendel used only pure-breeding plants.

....

.... **(2 marks)**

(b) Mendel pollinated the plants by hand.

....

.... **(2 marks)**

Guided

(c) Mendel repeated his plant crosses several times.

This allowed Mendel to collect .... of data.

It helped to reduce the effects of .... **(2 marks)**

2 Mendel carried out experiments with pea plants. In one experiment, he crossed a pure-breeding plant that produced yellow seeds with a pure-breeding plant that produced green seeds. He then crossed plants from the first generation with each other to produce a second generation. Mendel found that:

- all the first generation produced yellow seeds
- 349 second-generation plants produced yellow seeds
- 112 second-generation plants produced green seeds.

(a) Determine the percentage of plants in the second generation that produced yellow seeds.

....

....

....

Percentage = ... **(2 marks)**

(b) Mendel did not know about chromosomes or genes. Explain what he concluded from this experiment.

Mendel believed that inheritance of characteristics was determined by 'units'.

....

....

.... **(3 marks)**

3 Pea plants may be dwarf (short) or tall. Design an experiment that Mendel could have carried out to show that the unit for dwarf plants is recessive to the unit for tall plants.

Outline the important features of the method and the expected results. You do not need to suggest numbers of plants.

....

....

....

.... **(4 marks)**

Had a go ☐ Nearly there ☐ Nailed it! ☐

Fossils

1 Scientists may find preserved remains of woolly mammoths in frozen soil in the Arctic.

Guided

(a) Give **two** conditions that prevent the decay of dead organisms.

1 Low temperatures ....

2 .... **(2 marks)**

(b) Describe what is meant by a fossil.

You do not need to say how one forms.

....

.... **(2 marks)**

2 When people first settled on Christmas Island in the Pacific Ocean, scientists noted two species of rat that were unique to the island. In 1899, black rats escaped on to the island from a ship. These rats carried a parasite. The number of native rats decreased and by 1908 there were none left on the island.

Explain why the introduction of black rats led to the extinction of the native rats.

....

.... **(2 marks)**

3 Dinosaurs are thought to have become extinct around 65 million years ago, whereas mammoths became extinct about 4000 years ago. Modern humans evolved around 200 000 years ago.

(a) The extinction of the dinosaurs is often referred to as a mass extinction, in which all species became extinct. Give **three** causes of extinction, other than a single catastrophic event.

....

....

.... **(3 marks)**

(b) Explain whether it is likely that mammoths became extinct because of a single catastrophic event.

Use the information in the question to help you.

....

....

.... **(3 marks)**

4 Scientists study fossils to provide evidence for evolution in different species. Despite this research, scientists still cannot be certain about how life began on Earth. Explain why there are difficulties for scientists in studying the fossil record to provide evidence for evolution and for early life on Earth.

....

....

....

.... **(4 marks)**

Resistant bacteria

Guided

1 Evolution may be described as a change in the inherited characteristics of a population over time through a process of natural selection. Suggest whether evolution always results in a new species, and a give a reason for your answer.

No, because evolution can also produce new .... **(1 mark)**

2 MRSA is meticillin-resistant *Staphylococcus aureus*, a strain of bacterium. Meticillin is a similar substance to penicillin, used to treat bacterial infections.

(a) Name the type of substance meticillin is, and explain your answer.

Look at the information given in the question to help you.

....

.... **(2 marks)**

(b) Vancomycin is the most common and most effective antibiotic used in the treatment of MRSA infections. A new strain of MRSA is beginning to develop a resistance to vancomycin.

(i) Suggest why vancomycin-resistant MRSA is causing concern among doctors and scientists.

....

.... **(2 marks)**

(ii) Explain how populations of vancomycin-resistant MRSA developed from non-resistant populations of the bacteria.

You need to explain the increase in numbers of resistant bacteria and the decrease in numbers of non-resistant bacteria.

....

....

....

.... **(4 marks)**

3 Explain **one** way in which doctors or their patients can help to reduce the rate of development of antibiotic-resistant strains of bacteria.

....

.... **(2 marks)**

4 Explain how antibiotic resistance in bacteria provides evidence that supports Darwin's theory of evolution.

Think about the role of antibiotics in the development of antibiotic resistance.

....

....

.... **(3 marks)**

Had a go ☐ Nearly there ☐ Nailed it! ☐

Classification

Guided

1 The table shows how three different organisms are classified.

(a) Complete the table to give the correct classification group or binomial name for the organisms shown.

Classification group	Humans	Wolf	Panther
kingdom	Animalia	Animalia	Animalia
	Chordata	Chordata	Chordata
class	Mammalia	Mammalia	Mammalia
order	Primate	Carnivora	Carnivora
	Hominidae	Canidae	Felidae
genus	Homo	Canis	Panthera
	sapiens	lupus	pardus
binomial name	*Homo sapiens*		

(6 marks)

(b) Explain which **two** organisms in the table are most closely related.

Look for the lowest shared classification group.

....

.... **(2 marks)**

2 Virus classification begins at the level of order, not kingdom. Explain why scientists do not classify viruses in any kingdom.

....

.... **(2 marks)**

3 Honey badgers (*Mellivora capensis*), North American badgers (*Taxidea taxus*), Eurasian badgers (*Meles meles*), stink badgers (*Mydaus javanensis*) and ferret badgers (*Melogale personata*) are all members of the same mammalian family, but are otherwise not closely related to each other.

(a) Name the genus to which the ferret badger belongs.

.... **(1 mark)**

(b) Using evidence from the passage above, suggest an explanation for why the common names for the badgers are misleading.

....

.... **(2 marks)**

(c) Explain why a binomial naming system for each living organism is useful to scientists.

....

.... **(2 marks)**

Evolutionary trees

1 Bacteria and amoebae are unicellular (single-celled) organisms. However, bacteria are placed in the prokaryota kingdom and amoebae in the eukaryota kingdom.

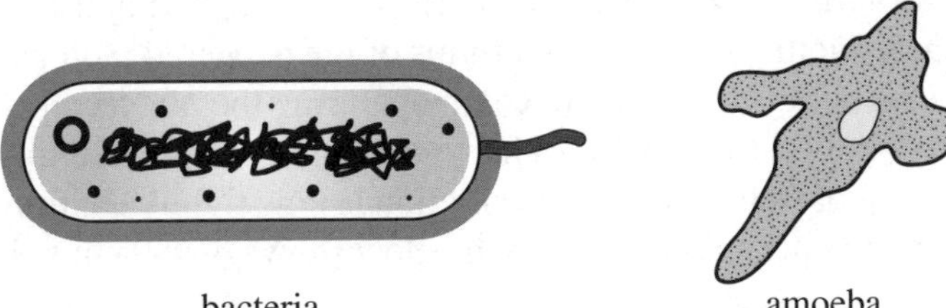

Suggest **one** reason why bacteria and amoebae were placed in different kingdoms.

> Think about the differing features of bacterial cells and animal cells.

....

.... **(2 marks)**

Guided

2 Carl Woese proposed that organisms should be classified into three domains. Complete the table to give the missing information.

> There is one type of organism in the archaea domain and four in the eukaryota domain.

Domain	Type(s) of organism in domain
archaea	
bacteria	true bacteria and cyanobacteria
eukaryota	protists,

(6 marks)

3 The diagram shows an evolutionary tree.

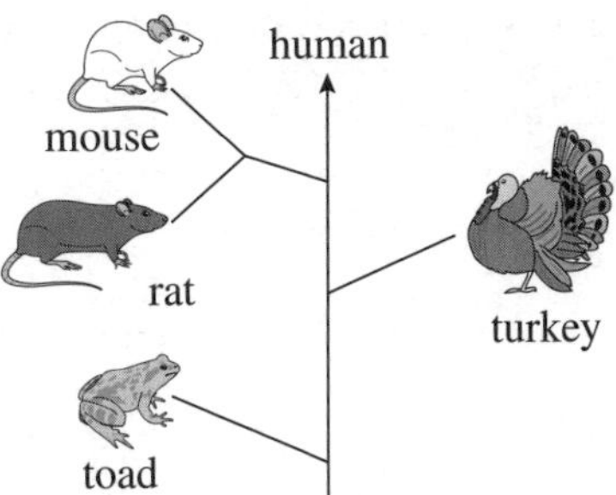

(a) Identify the organism that shares the oldest common ancestor with humans.

.... **(1 mark)**

(b) Give the names of the **two** organisms on this tree that are most closely related.

.... **(1 mark)**

(c) Mice and rats are both classified in the order Rodentia. Use the evolutionary tree to justify this decision.

....

.... **(2 marks)**

Had a go ☐ Nearly there ☐ Nailed it! ☐

Extended response – Inheritance, variation and evolution

The peppered moth, *Biston betularia*, rests on the trunks of trees during the daytime. Some species of birds eat these moths. There are two forms of the peppered moth, a dark form and a light form. Dark moths became much more common during the nineteenth century. One theory suggested that pollution from factories had blackened the light tree trunks with soot. Birds could not easily see the dark moths. Air pollution is now tightly controlled by law and soot emissions have sharply decreased. The graph below shows how the numbers of dark moths has changed.

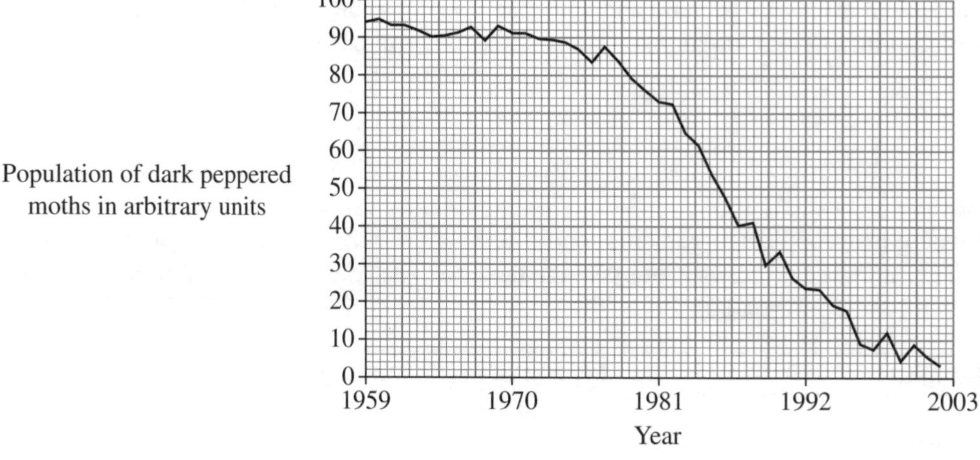

Explain the change in the number of dark moths using Darwin's theory of evolution by natural selection.

It may help if you make a brief plan before you start writing. You may wish to consider:

- What did the moth population look like in the mid-twentieth century?
- How has the moth population changed over time? Use data from the graph to help you.
- What does the theory of evolution say, and how does it relate to changes in *Biston betularia*?

....

....

....

....

....

....

....

....

....

....

.... **(6 marks)**

Ecosystems

1 Light intensity is an abiotic factor in a community.

(a) Plants may compete with each other for light. Give **two** other abiotic factors that plants may compete for.

.... **(2 marks)**

Guided

(b) Give **two** things that animals may compete for.

These could be biotic factors and/or abiotic factors.

territory and ....

.... **(2 marks)**

2 Draw **one** line from each term to its meaning.

Term	Meaning
community	a single living individual
organism	all the living organisms and non-living parts in an area
population	all the different populations in a habitat
ecosystem	all the organisms of the same species in a habitat

(4 marks)

3 Lichens are organisms that consist of a fungus closely associated with an alga. The fungus can digest plant material and absorb nutrients, and the alga can make food by photosynthesis. Lichens can grow on the trunks of trees and on other surfaces.

Scientists looked at the distribution of lichens on a tree and on a concrete post next to the tree. The bar chart shows the results.

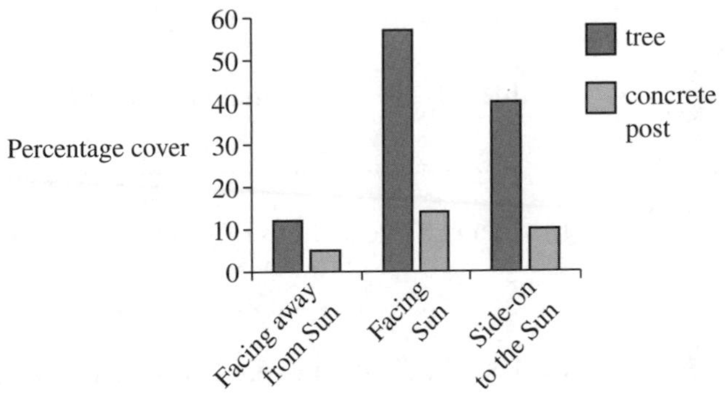

(a) Describe what the results show.

What are the differences between the results for the tree and the post, and for the different sides?

....

....

.... **(2 marks)**

(b) Explain the results.

....

....

....

.... **(4 marks)**

Had a go ☐ Nearly there ☐ Nailed it! ☐

Interdependence

Guided

1 Give the meaning of a stable community.

All the species and .... are in balance so that

the population sizes .... (2 marks)

2 The diagram shows an interdependent community of plants and animals.

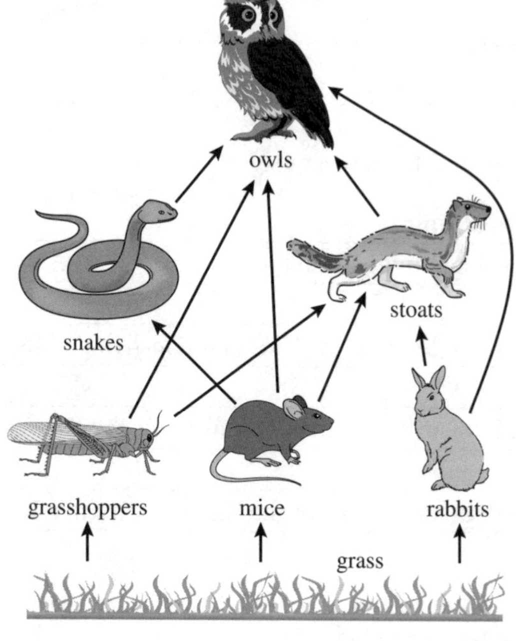

(a) Identify the animal that has the most sources of food in the community.

.... (1 mark)

(b) Identify **two** sources of food for stoats in the community.

.... (2 marks)

(c) Use the diagram to predict the effect on the number of snakes if the number of mice decreased. Explain your answer.

....

.... (2 marks)

(d) Use the diagram to predict the effect on the number of mice if all the owls were removed from the community. Explain your answer.

> Work through all the feeding relationships that lead from mice to owls. Explain clearly how and why the number of mice is affected by the removal of the owls.

....

....

....

.... (3 marks)

Adaptation

1 Hydrothermal vents are found on the seabed. Water that comes out of these vents is very hot. A species of bacteria survives in this hot water. Several other different organisms are found nearby.

(a) Give the name used for organisms, such as these bacteria, that are adapted to survive in hostile conditions.

.... **(1 mark)**

(b) The seabed is cold and dark. There are usually few organisms there. Explain why there is a large variety of life around hydrothermal vents, but not elsewhere on the seabed.

Think about the type of organism that usually produces food that other organisms can consume.

....

....

.... **(3 marks)**

2 Many plants, such as roses, need to attract insects so that the plants can be pollinated and produce seeds.

(a) Suggest how these plants are adapted in order to attract insects.

Guided

Roses produce flowers to attract insects. These flowers have very bright

...., and they also give off a strong .... **(2 marks)**

(b) The diagram shows the seeds that are produced by dandelions. Suggest how these seeds are adapted to make sure that they can be dispersed.

....

....

....

.... **(2 marks)**

3 The diagram shows two different species of animal, A and B.

dark skin on face

thick coat

wide feet

Animal A

large ears

thin coat

Animal B

Justify means that you need to use evidence from the information supplied to support your answer.

Determine which animal, A or B, would be better adapted to survive in a cold environment. Justify your answer.

....

....

.... **(3 marks)**

Had a go ☐ Nearly there ☐ Nailed it! ☐

Food chains

1 The diagram shows an interdependent community of plants and animals.

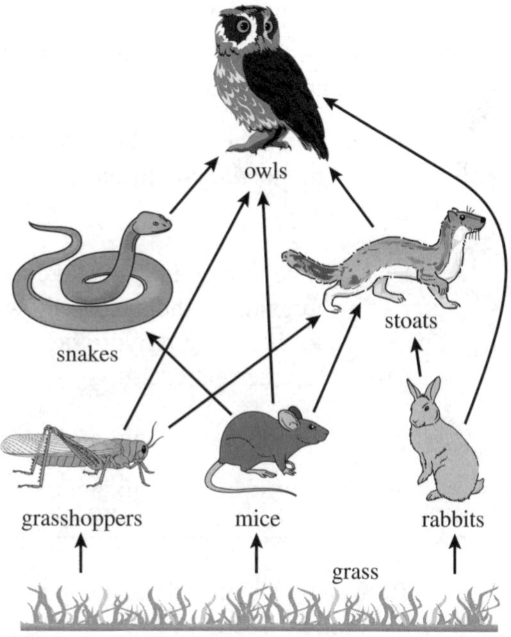

Make sure that your answers name organisms from the community shown in the diagram.

(a) Identify **one**:

(i) primary consumer:

.... **(1 mark)**

(ii) tertiary consumer:

.... **(1 mark)**

(b) Write a food chain that involves snakes.

....

.... **(1 mark)**

(c) Describe the function of the grass in this food chain: grass → rabbit → owl.

....

.... **(2 marks)**

Guided

2 Aphids are insects that feed on plant sap in phloem. Ladybirds are insects that eat aphids.

The graph shows how the population sizes of aphids and ladybirds in a community change over time.

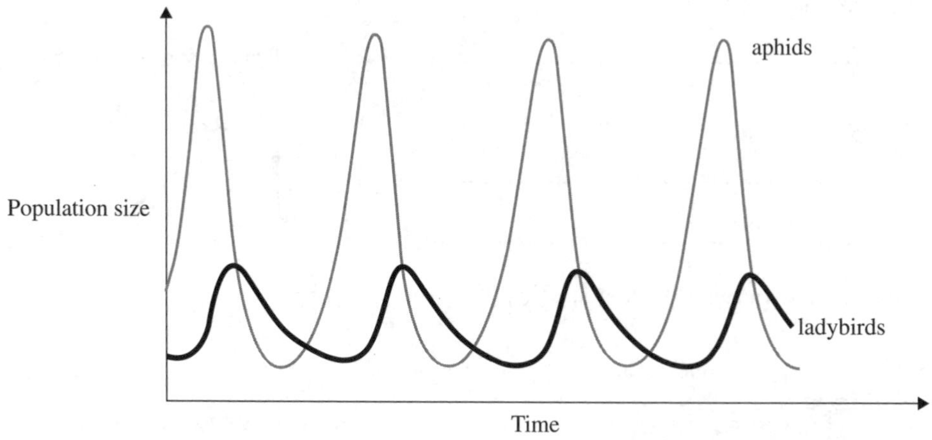

Explain the changes seen in the graph.

You need to give reasons for a rise and fall in each population.

The number of aphids rises if there is plenty of food and few ladybirds, because more survive to reproduce. More aphids mean more food for the ....

....

....

.... **(4 marks)**

Fieldwork techniques

1 A gardener went into his garden every night at 7 pm. He counted the number of slugs in the same 1 m^2 area of his flower bed. The table shows his results.

Day	1	2	3	4	5	6	7
Number of slugs	11	12	7	12	8	8	12

(a) Calculate the mean number of slugs seen each day.

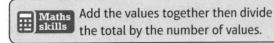

Add the values together then divide the total by the number of values.

....

....

Mean number of slugs = ... **(1 mark)**

(b) Determine the median number of slugs.

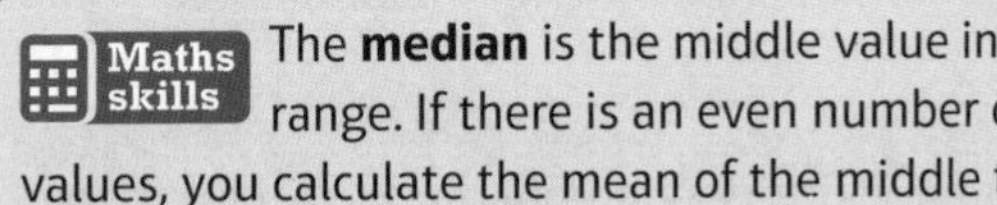

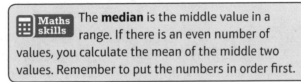

The **median** is the middle value in a range. If there is an even number of values, you calculate the mean of the middle two values. Remember to put the numbers in order first.

Median number of slugs = ... **(1 mark)**

(c) Give **two** reasons why a quadrat would be suitable to use in this investigation.

....

.... **(2 marks)**

2 A species of broad-leaved plants is growing in a small field between a path and a woodland. Describe how you would use a belt transect to investigate the distribution of this species growing between the path and the woodland.

Make sure that you describe the use of quadrats, the measurements you would take and what you would record.

....

....

.... **(3 marks)**

3 A group of students investigated the number of clover plants on a football pitch. The pitch measured 100 m by 65 m. The students used a 1 m × 1 m quadrat. They found that the mean number of clover plants in each quadrat was 7. Estimate the number of clover plants on the whole football pitch.

Area of the football pitch = 100 × 65 = ... m^2

Number of clover plants = 7 × ....

= .... **(2 marks)**

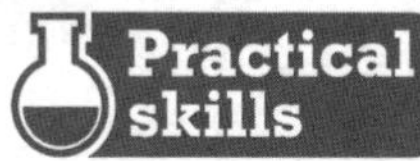

Field investigations

1 A student surveyed the distribution of a species of lichen growing on the trunk of a tree. He used a small quadrat to measure the percentage cover by these lichens on the south-facing and north-facing sides of the tree. He used a light meter to measure the light intensity on each side. The table shows his results.

	South-facing side	North-facing side
Mean percentage cover	39	4
Light intensity in units	320	228

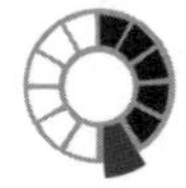

(a) The student concluded that the lichen was better adapted to growing in areas with higher light intensities. Explain whether his conclusion was correct.

You are asked to say only whether he was right or not, and why. Do not try to explain his results.

....

.... **(2 marks)**

(b) Light intensity is an abiotic factor. Suggest one other abiotic factor that might be responsible for the different distribution of lichen on the tree trunk, and give a reason for your answer.

....

.... **(2 marks)**

2 Limpets are animals that have a shell. They live on rocks that are underwater for some or all of the time. They can be found attached to rocks on the seashore.

A scientist investigated the distribution of limpets on a beach. She set up three different transects, and measured the number of limpets inside quadrats placed at regular intervals. The table shows her results.

Distance from the sea in m	Number of limpets			Mean number of limpets
	Transect 1	Transect 2	Transect 3	
0.5	20	23	20	21
1.0	18	16	17	17
1.5	13	13	13	13
2.0	10	8	9	9
2.5	5	6	4	5
3.0	1	2	0	1

(a) Determine the median number of limpets in Transect 2.

Maths skills The median is the middle value in a range. If there is an even number of values, you calculate the mean of the middle two values.

Median number of limpets in Transect 2 = ... **(1 mark)**

Guided

(b) Describe a suitable conclusion that can be made from the results of this investigation.

What pattern can you see, and what does it mean for the survival of limpets on the seashore?

As the distance from the sea increases, the number of limpets ....

....

.... **(3 marks)**

Cycling materials

Guided

1 Complete the diagram of the carbon cycle by writing the names of the processes in the boxes.

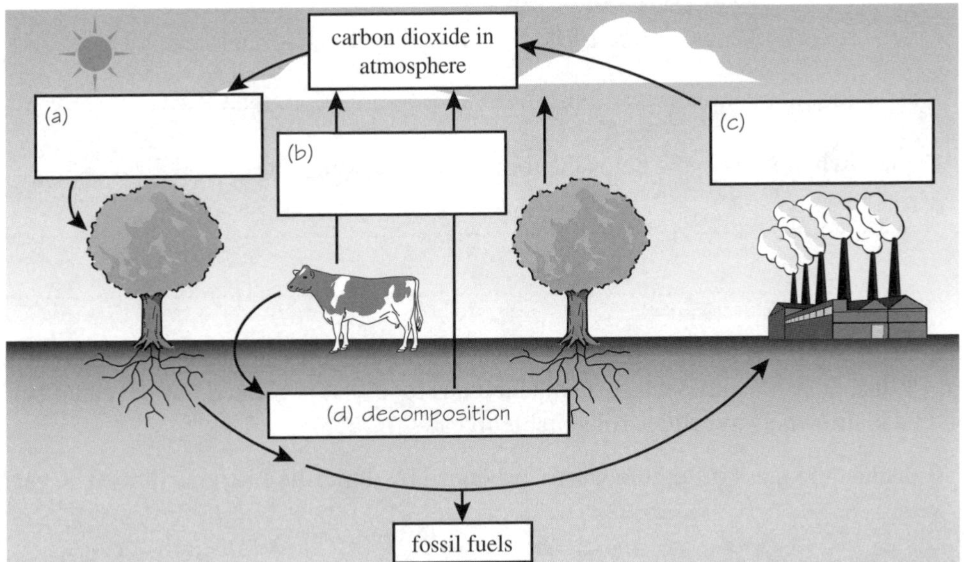

(4 marks)

Guided

2 In parts of California there is often a lack of rainfall. Water is taken from rivers and used to water lawns and golf courses. When there is a drought, restrictions are placed on the number of days each week that golf courses can be watered. Suggest an explanation for restricting water use.

Think about the process involved in the water cycle, which provides plants and animals with fresh water.

Water vapour escapes from plants by .... and from rivers and soil by .... If the water is not replaced by rainfall, ...

.... **(3 marks)**

3 Explain why microorganisms are important in recycling carbon in the environment.

....

.... **(2 marks)**

4 The diagram shows a community of plants and fish in a fish tank. Explain how carbon is recycled between these organisms.

You do not need to describe the role of microorganisms here, but you must mention carbon dioxide.

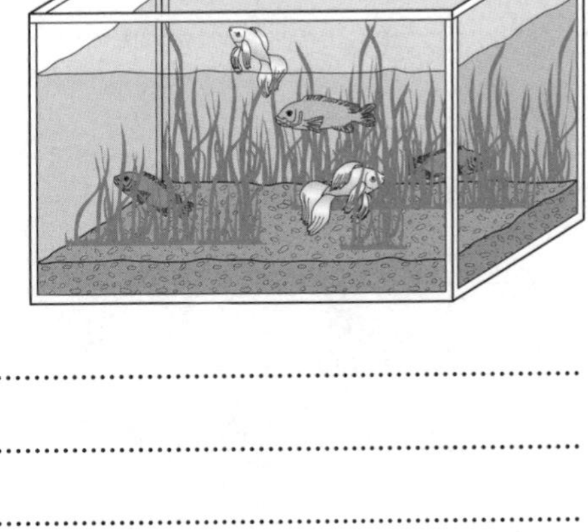

....

....

....

.... **(4 marks)**

Had a go ☐ Nearly there ☐ Nailed it! ☐

Decomposition

1 Gardeners may grow tomatoes in compost. Compost can be made from garden waste.

(a) Give the conditions needed for compost to form from garden waste.

Guided

Microorganisms that decay the garden waste need aerobic conditions, and

.... and .... **(3 marks)**

(b) Explain why the compost helps the tomatoes grow well.

....

.... **(2 marks)**

2 In the UK, 7.5 million tonnes of kitchen waste are produced by domestic kitchens each year. Of this, 60% is 'edible waste'. This includes food that is cooked but not eaten, and food that is thrown away while still suitable to eat.

(a) Calculate the mass of 'edible waste' produced by domestic kitchens in the UK each year.

....

....

Mass = ... million tonnes **(2 marks)**

(b) Large amounts of waste are disposed of in landfill sites. The waste is buried then covered with soil. A typical landfill site can process 2500 tonnes of waste material per day.

(i) Describe what is meant by anaerobic conditions.

.... **(1 mark)**

(ii) Name the flammable gas produced by the anaerobic decay of waste food.

.... **(1 mark)**

(iii) Suggest a use for the gas named in part (ii).

.... **(1 mark)**

3 Food can be preserved from aerobic decay in various ways. These include drying, refrigeration and packing the food in nitrogen. Explain how these methods help to preserve food.

Drying: ....

....

Refrigeration: ....

....

Packing in nitrogen ....

....

Nitrogen is the unreactive gas that makes up 78% of the atmosphere.

.... **(6 marks)**

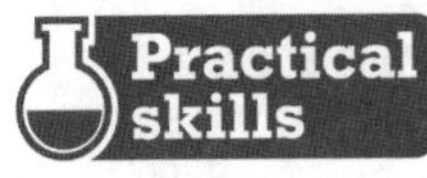

Investigating decay

1 Some scientists investigated the decomposition of leaves from a single tree. They placed leaves in two mesh bags with different mesh sizes. The bags were buried next to each other in the soil, then removed every 4 weeks. The mass of the leaves inside was measured each time. The table shows their results.

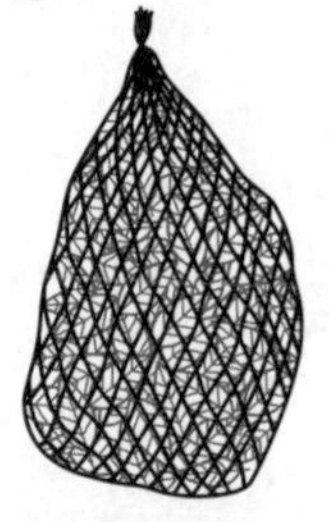
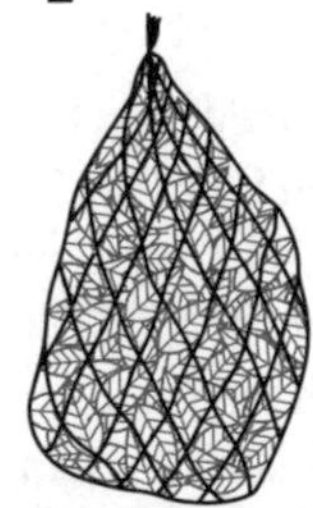

		0 weeks	4 weeks	8 weeks	12 weeks	16 weeks	20 weeks
Mass of leaves as % of starting mass	Small mesh	100	92	84	74	72	68
	Large mesh	100	88	78	66	60	58

(a) Identify **two** ways in which the scientists designed their experiment to obtain valid results.

Guided

The scientists buried the bags next to each other, so the conditions would be the same. They also used .... **(2 marks)**

(b) Explain the results of this experiment.

Think about what can get in and out of the bags.

....

....

.... **(3 marks)**

2 A large commercial biogas generator consumes garden waste and food waste. The graph shows how the temperature inside the generator affects the rate of biogas production.

Biogas production in m^3/day (0 to 4.0) against Temperature in °C (0 to 50)

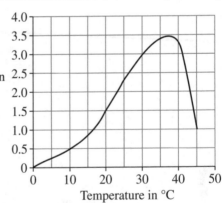

(a) Determine the temperature at which the rate of biogas production is highest.

.... **(1 mark)**

(b) Explain the changes in the rate of biogas generation seen in the graph.

Describe the changes seen and give reasons for them. Remember that enzymes are involved in decay.

....

....

....

.... **(4 marks)**

Had a go ☐ Nearly there ☐ Nailed it! ☐

Environmental change

1 Turtles lay their eggs in sand on beaches. The temperature of the sand can affect the sex of the turtles that hatch from the eggs. The graph shows this relationship.

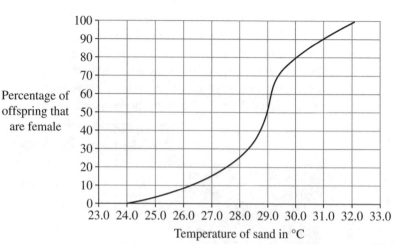

Percentage of offspring that are female

100, 90, 80, 70, 60, 50, 40, 30, 20, 10, 0

23.0 24.0 25.0 26.0 27.0 28.0 29.0 30.0 31.0 32.0 33.0

Temperature of sand in °C

(a) Use the graph to estimate the temperature that would give equal numbers of male and female turtles.

....

.... **(1 mark)**

(b) A turtle lays 120 eggs in the sand. Calculate the effect on the number of males hatching from the eggs if the temperature increases from 30 °C to 31 °C.

Maths skills Use the graph to work out the percentage of males hatching at each temperature, then use the difference.

....

.... **(2 marks)**

2 The graph shows the mass of carbon dioxide absorbed by a tree during the year.

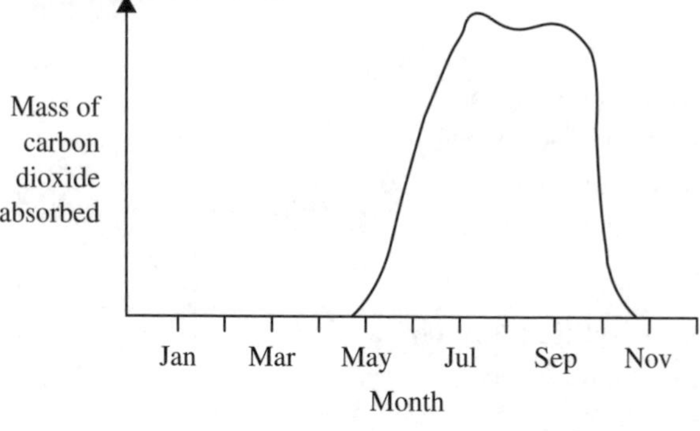

(a) The tree absorbs carbon dioxide for photosynthesis. The concentration of carbon in the atmosphere is a factor that affects the rate of photosynthesis. Give **three** other factors that affect its rate.

amount of chlorophyll, ... *and* .... **(3 marks)**

(b) Explain the changes seen in the graph.

Think about how the tree and the environmental conditions change during the year.

....

....

....

.... **(4 marks)**

Waste management

1 Sulfur dioxide is a major cause of acid rain. It is produced by some industrial processes and by the use of some fossil fuels.

(a) Describe what is meant by pollution.

Guided

Pollution is the release or presence in the environment of ....

.... **(1 mark)**

(b) A scientist investigated the effects of sulfur dioxide pollution on living organisms. She studied the pattern of growth of a species of lichen on the bricks of two walls. One wall was in an area with air polluted by sulfur dioxide, and the other in an area with unpolluted air. The diagram shows these two walls.

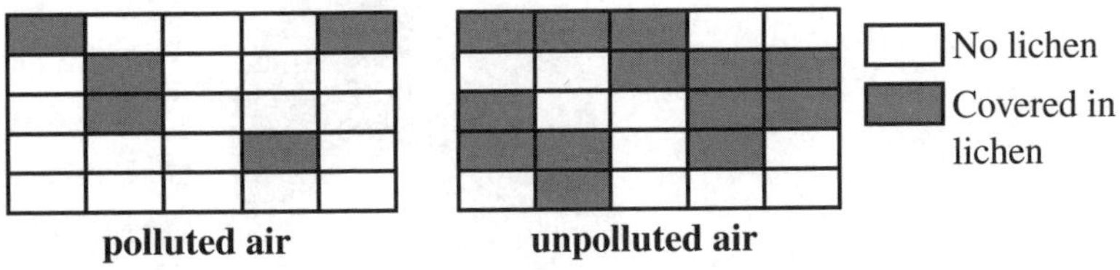

Describe what the results show.

....

.... **(2 marks)**

2 The rapid increase in the world's human population has caused an increased use of resources.

(a) Give **one** other reason why the use of resources has increased over time.

.... **(1 mark)**

(b) Describe **two** ways in which human activities can lead to reduced space for animals and plants.

....

.... **(2 marks)**

3 The overuse of artificial fertilisers can cause pollution in water. Water plants absorb mineral ions from the fertilisers and grow very quickly. They cover the surface of the water and prevent light reaching plants below. These plants die and are decomposed. The decomposer organisms use dissolved oxygen to respire.

(a) Predict the effect on oxygen levels in river water polluted by artificial fertilisers. Justify your answer.

> **Justify** means that you should use evidence from the information supplied to support your answer.

....

....

.... **(3 marks)**

(b) The number of fish can decrease if river water is polluted by artificial fertilisers. Suggest **two** reasons that explain why.

....

....

.... **(3 marks)**

Had a go ☐ Nearly there ☐ Nailed it! ☐

Deforestation

1 Describe what is meant by deforestation.

.... (1 mark)

2 Rainforests in Indonesia are being cut down. The pie chart shows how the deforested land is used.

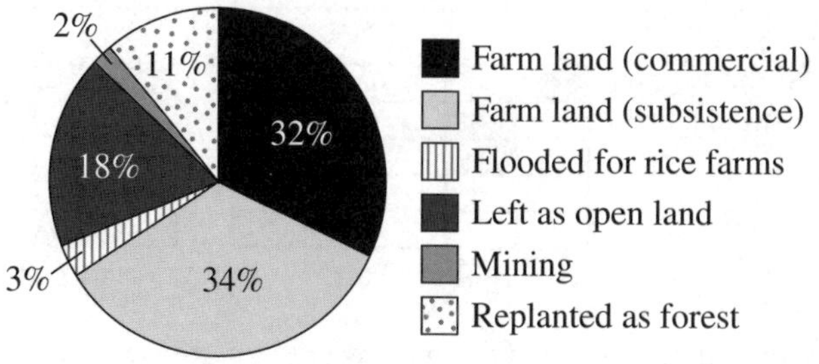

Use of deforested land	Percentage
Farm land (commercial)	32%
Farm land (subsistence)	34%
Flooded for rice farms	3%
Left as open land	18%
Mining	2%
Replanted as forest	11%

(a) Calculate the total percentage of land that is used for farming after deforestation.

Percentage = ... (1 mark)

(b) Suggest **one** use for the trees that are cut down in rainforests.

.... (1 mark)

Guided

(c) Of the land that is cleared 11% is replanted as forest. Replanting some trees is important in terms of controlling the levels of carbon dioxide in the atmosphere. Explain why.

Planting trees .... the rate at which carbon dioxide is taken from the atmosphere. The carbon dioxide is 'locked up' as.................................. (2 marks)

3 Deforestation often leads to a loss of biodiversity.

(a) Describe what is meant by biodiversity.

.... (1 mark)

(b) Describe why deforestation can lead to a loss of biodiversity.

....

.... (2 marks)

4 Compost is used by many gardeners for growing plants from seeds. Some compost sold in the UK contains peat. Suggest the effect on the environment of using peat-based composts, and explain your answer.

> Make sure that you answer in terms of the environment, not in terms of the effects on gardening.

....

....

.... (3 marks)

Global warming

1 Scientists drill into the ice near the Vostok research station in Antarctica. They analyse the ice and ancient air trapped in it. This lets them determine average world temperatures and atmospheric carbon dioxide levels in the distant past. The graph shows their results.

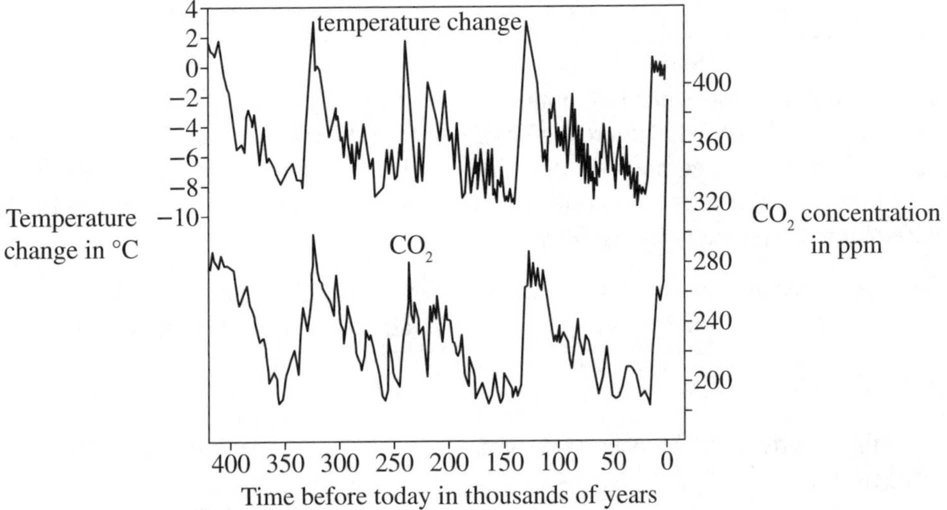

(a) Describe the relationship between average world temperatures and carbon dioxide levels, as shown by the graph.

A positive gradient in the temperature change line shows that the average world temperature increased.

....

....

.... **(3 marks)**

(b) Explain a possible consequence of the most recent changes in carbon dioxide levels for average temperatures.

....

.... **(2 marks)**

2 Carbon dioxide is a greenhouse gas. Increased levels of this gas in the atmosphere contribute to global warming.

(a) A different greenhouse gas is released because of human activities such as rice farming and cattle farming. Name this gas.

.... **(1 mark)**

(b) Describe what is meant by global warming.

Guided

an increase in the Earth's average ....

due to rising levels of .... **(2 marks)**

3 Global warming may cause climate change and increases in sea levels. Describe **two** biological consequences of global warming.

Make sure you answer in terms of its effects on living things.

....

.... **(2 marks)**

Had a go ☐ Nearly there ☐ Nailed it! ☐

Maintaining biodiversity

1 Give **two** reasons, other than to ensure the stability of an ecosystem, why it is important to maintain biodiversity.

1 ....

2 .... **(2 marks)**

2 Hedgerows were originally planted to mark the boundaries of fields or to stop farm animals escaping. A typical hedgerow contains several different species of shrubs, and may also contain trees. Beginning in the middle of the last century, hedgerows were removed to form larger fields. It is estimated that the total length of hedgerows in England decreased by more than half in just 50 years.

(a) Suggest a reason why farmers wanted to increase the size of their fields.

> Think about the differences between modern farming and traditional farming.

.... **(1 mark)**

(b) Explain the effect on biodiversity of removing hedgerows.

> Do not just answer in terms of the hedgerow plants themselves.

....

....

.... **(3 marks)**

(c) Suggest a benefit of replanting hedgerows:

(i) to farmers

.... **(1 mark)**

(ii) to biodiversity

.... **(1 mark)**

3 The area of woodland in the UK has increased from less than 5% of the land surface just over 100 years ago, to over 12% today. Describe **two** benefits for the environment of reforestation such as this.

....

.... **(2 marks)**

4 Hedgehogs roam up to 2 km each night, looking for food. Explain why building small tunnels under new roads may help to conserve the numbers of hedgehogs.

Guided

Hedgehogs may be run over by traffic if they ....

....

.... **(3 marks)**

Trophic levels

1 Deer are animals that feed on grass and leaves. They are a component of trophic level 2 in their food chain.

(a) Give **two** terms that describe a species in this position in a food chain.

.... **(2 marks)**

(b) Describe the role of the grass in a food chain containing grass and deer.

Guided

The grass is a plant, so it is a .... which makes

.... **(2 marks)**

2 The table shows some information about organisms in a food chain.

Guided

Organism	Number of organisms	Mass of each organism in g	Total biomass in kg
clover plant	10 000	5	50
mouse	600	25	
snake	10	500	

Complete the table to show the total biomass for the organisms in this food chain.

Maths skills: Remember to convert from g to kg in your answer.

(3 marks)

3 The table shows the total biomass at each trophic level in a food chain.

Organism	Total biomass in kg
producer	2500
herbivore	500
carnivore	200

Maths skills: Use the data in the table. Draw your diagram to scale using a ruler, and label each trophic level.

(a) Draw a pyramid of biomass for this food chain.

(2 marks)

(b) Describe **two** reasons why only some of the biomass is transferred from one trophic level to the next.

1 ....

....

2 ....

.... **(4 marks)**

Had a go ☐ Nearly there ☐ Nailed it! ☐

Food security

1 Sustainable methods of food production must be found to feed everyone on Earth.

(a) Give the meaning of food security.

.... **(1 mark)**

(b) Describe what it means if a process is sustainable.

Guided

It meets the needs of people today, without reducing ....

.... **(2 marks)**

2 The cost of food affects how easily people can afford their diet. For example, farmers need to buy tractors and other farm machinery. The food they produce is likely to cost more if these are expensive. Give **two** other factors that can affect the cost of food.

....

.... **(2 marks)**

3 Human activities may threaten food security.

(a) Describe how wars in some parts of the world affect food security there.

....

.... **(2 marks)**

(b) Explain how an increasing human birth rate may affect the level of food security.

....

.... **(2 marks)**

4 There are different ways to manage pests.

(a) Complete the table to suggest how different pest management strategies work.

Pest management strategy	How it controls pests
introduce a natural predator of the pest	
spray insecticides	

(2 marks)

(b) Explain how managing pests helps to improve food security.

Think about the effect of pests on crop yields or the quality of food produced.

....

.... **(2 marks)**

5 Bioethanol is mixed with petrol for use in cars. Bioethanol is produced by the fermentation of plant sugars. These sugars are obtained from crop plants such as wheat, maize and sugar cane. Explain how bioethanol production may affect the food supply for human populations.

....

....

.... **(3 marks)**

Farming techniques

1 A herd of cows grazes in a field of grass. Every 1 m^2 of the field receives an average of 1240 MJ of light energy per year. On average, each 1 m^2 of growing grass transfers 14 900 kJ of energy to its energy stores.

(a) Calculate the percentage of light energy that is transferred to energy stores in the grass.

Maths skills Remember to convert the units so that they are the same: 1 MJ = 1000 kJ

....

....

Percentage = ... % **(2 marks)**

(b) The diagram shows the energy transfers occurring when a cow eats food. It does not show the energy stored in meat and other parts of the cow.

540 kJ transferred to surroundings

150 kJ lost in urine and faeces

1000 kJ in food

(i) Calculate the amount of energy stored in meat and other parts of the cow.

....

....

Energy stored in cow = ... kJ **(1 mark)**

(ii) Give **two** ways, other than being lost in urine and faeces, in which energy may be transferred to the surroundings by the cow.

Guided

keeping warm and .... **(2 marks)**

(c) Use the information in part (b) to explain why some people think it is better to produce crops for food, rather than producing meat to eat.

....

....

.... **(3 marks)**

2 A student investigated one aspect of farming chickens indoors and outdoors. He placed a beaker of hot water on the bench to represent a chicken living outdoors. The student placed another beaker, containing the same volume of hot water, inside a cardboard box to represent a chicken living indoors in a small cage. He then measured the temperature of the water in each beaker over a 30-minute period. The graph shows his results.

Temperature in °C (40–70); Time from start of experiment in minutes (0, 10, 20, 30); beaker in box; beaker on bench

Explain, using these results and your own knowledge, why the farming of chickens indoors may be more efficient than the farming of chickens outdoors.

....

....

....

.... **(4 marks)**

Sustainable fisheries

1 Fishing boats have strict quotas in the North Sea. The quotas limit the numbers and types of fish that can be caught and brought ashore.

(a) Describe how a quota system helps in the management of fish stocks.

....

.... **(2 marks)**

(b) Explain how the success of the quota system would be affected if it did not apply to all fishing boats.

....

.... **(2 marks)**

2 Salmon intended for human consumption are caught in the wild or are farmed. The pie charts show the proportions of salmon produced in these ways. The data are shown for two different years.

1990: 25% farmed; 2015: 30% wild

Key: ■ Wild ☐ Farmed

(a) Describe what the pie charts show, using the data to support your answer.

> **Maths skills** Include numbers in your answer.

....

.... **(2 marks)**

Guided

(b) Farmed salmon are kept in large numbers in cages in the sea. Salmon are mostly carnivorous and are fed on pellets of fish food. Farmed salmon are often higher in fat content than wild salmon. Use the information above to suggest why.

Farmed salmon are kept in cages, so ....

As they are provided with food,.. **(2 marks)**

(c) As humans eat more farmed salmon, there is an effect on wild salmon and other fish. Explain how farming salmon could change the numbers of wild salmon and other fish.

....

....

.... **(3 marks)**

3 The mesh size of a fishing net is the size of the spacing between its individual strands. Explain how the mesh size can contribute to the conservation of wild fish stocks.

> Think about how the right sort of net can select the species of fish caught, and the differences between fish that are caught and those that are not.

....

....

....

.... **(4 marks)**

Biotechnology and food

1 Mycoprotein is produced by a fungus called *Fusarium*. Mycoprotein can be made into a variety of foods, such as a burger. The table shows some nutritional information about mycoprotein, lean beef and lentils.

	Mass of nutrient (in 100g of food) in g		
Food	**Carbohydrate**	**Protein**	**Fat**
mycoprotein	2	14	3
lean beef	0	25	9
lentils	17	9	1

(a) Identify the food with the highest protein content.

.... **(1 mark)**

(b) Explain why mycoprotein burgers would be suitable for vegetarians.

Think about the nutrient content of these burgers and where they come from.

....

.... **(2 marks)**

2 The diagram shows a fermenter, which is used to grow large quantities of *Fusarium*.

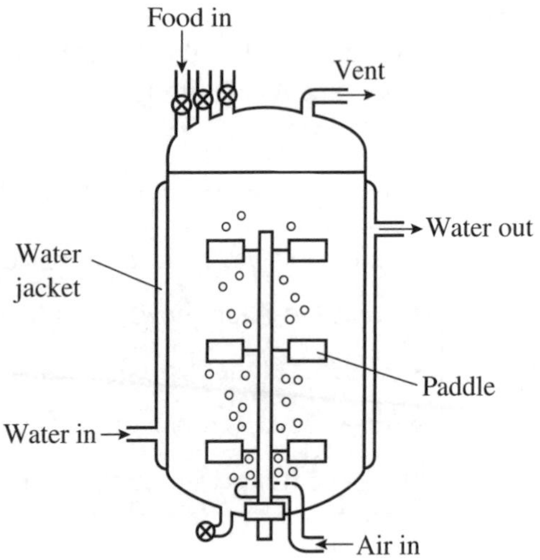

(a) Name the food provided to *Fusarium* in the fermenter.

....

....

(1 mark)

(b) Explain why air is fed into the fermenter.

Guided

Air contains ....

It is supplied because Fusarium must grow in

.... conditions. **(2 marks)**

(c) Ammonia is supplied to the fermenter. Suggest a reason to explain why.

....

.... **(2 marks)**

(d) *Fusarium* is a living organism. Suggest why it is important to have a cold water jacket around the fermenter.

....

....

.... **(3 marks)**

Had a go ☐ Nearly there ☐ Nailed it! ☐

Extended response – Ecology

The graph shows how the area of land being deforested in Brazil changed between 1992 and 2009.

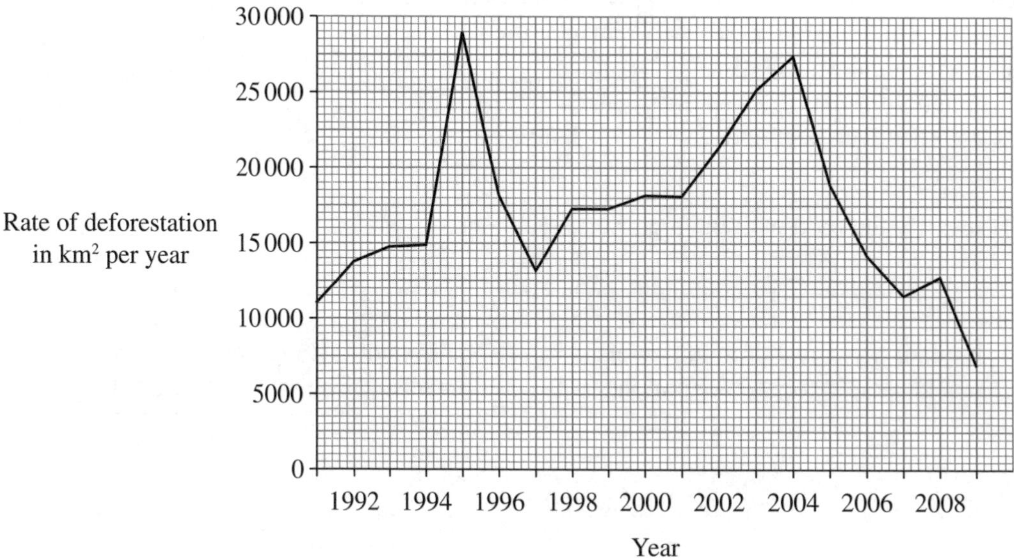

It is estimated that about half the world's tropical forests have been cleared for farmland and timber. The remaining forests are estimated to store around 280 billion tonnes of carbon in their biomass.

Explain how deforestation can contribute to climate change.

> It may help if you make a brief plan before you start writing. You may wish to consider:
> - what the graph shows about how the area of forest has changed over time
> - the uses of the cleared land and the wood from the felled trees
> - relevant parts of the carbon cycle.
>
> Think about how climate change may be linked to these points.

....

....

....

....

....

....

....

....

....

....

....

.... **(6 marks)**

Timed Test 1

Time allowed: 1 hour 45 minutes

Total marks: 100

AQA publishes official Sample Assessment Material on its website. This practice exam paper has been written to help you practise what you have learned and may not be representative of a real exam paper.

1 The diagram shows a bacterial cell and a plant cell.

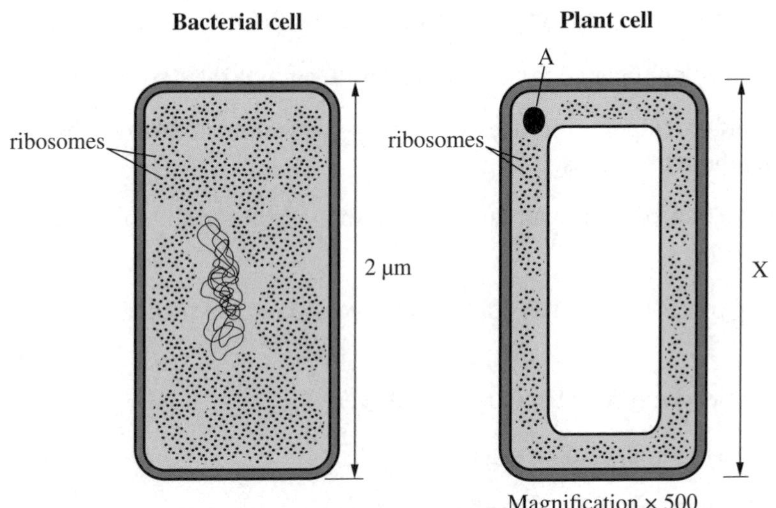

Magnification × 500

(a) What is the structure labelled **A** in the diagram? Tick **one** box.

mitochondrion ☐
vacuole ☐
nucleus ☐
chloroplast ☐ **(1 mark)**

(b) Both types of cell contain ribosomes. Give the function of a ribosome. **(1 mark)**

(c) Describe **two** ways in which bacterial cells and plant cells differ in their structure. **(2 marks)**

(d) Although the cells are drawn the same size, the magnifications used are different. The actual length of the bacterial cell is 2 μm. Calculate the actual length, **X**, of the plant cell in metres. Give your answer in standard form and to 2 significant figures. Show your working. **(3 marks)**

2 Uncontaminated cultures of bacteria are needed in order to carry out research into antibiotics.

(a) Explain why Petri dishes and culture media must be sterilised before use. **(2 marks)**

(b) Describe how inoculating loops, used to transfer bacteria to the culture media, are sterilised. **(1 mark)**

(c) Describe how the lid of the Petri dish must be secured. **(2 marks)**

(d) Explain why, in school laboratories, cultures should generally be incubated at 25 °C. **(2 marks)**

3 A student tested the effect of the type of milk on bacterial growth. Resazurin can be used to show the growth of a population of microorganisms. Resazurin is a blue dye but it becomes colourless if there are many bacteria present.

The student placed some pasteurised milk in a test tube. She placed sterilised milk in another tube. She put the tubes in a water bath set at 36 °C and incubated them for 1 day. Then she added blue resazurin dye to each tube. One hour later she recorded the colour. The table shows her results.

Type of milk	Resazurin colour after 1 hour
pasteurised	colourless
sterilised	blue

(a) Suggest a reason why the student used a water bath. **(1 mark)**

(b) The tube containing sterilised milk acted as a control. Give the function of this control. **(1 mark)**

(c) Explain why there was a colour change for pasteurised milk but not for sterilised milk. **(3 marks)**

4* Diffusion and active transport are important processes in plants and animals. Many substances move into or out of cells by these processes.

Compare diffusion and active transport. In your answer, you should refer to suitable examples of these processes in plant or animal cells. **(6 marks)**

5 In a developing human embryo, tissues such as epithelial tissue and organs such as the liver develop from stem cells.

(a) Name the process by which stem cells become specialised. **(1 mark)**

(b) Describe what is meant by a tissue and an organ. **(3 marks)**

(c) Give **one** way in which adult stem cells differ from embryonic stem cells. **(1 mark)**

6 The digestive system is an organ system in which organs work together to digest and absorb food.

(a) Which statement about bile is correct? Tick **one** box.

It is made in the gall bladder and stored in the liver. ☐
It is made in the pancreas and stored in the gall bladder. ☐
It is made in the liver and stored in the gall bladder. ☐
It is made in the gall bladder and stored in the pancreas. ☐ **(1 mark)**

(b) Describe the functions of bile and lipase in the digestion of lipids to form soluble products. **(4 marks)**

(c) The diagram shows a villus, a structure that lines the wall of the small intestine.

Explain **two** ways in which the structure of a villus is adapted for its function in the digestive system. **(4 marks)**

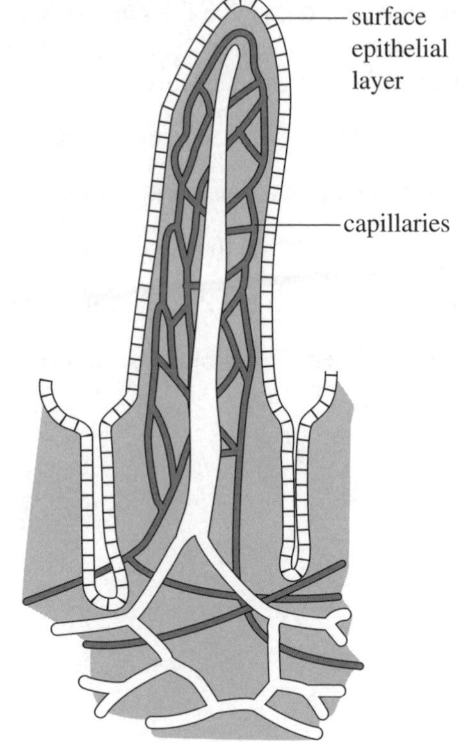

(d) Isomerase is an enzyme used in the food industry. The graph shows the relative activity of isomerase at different temperatures.

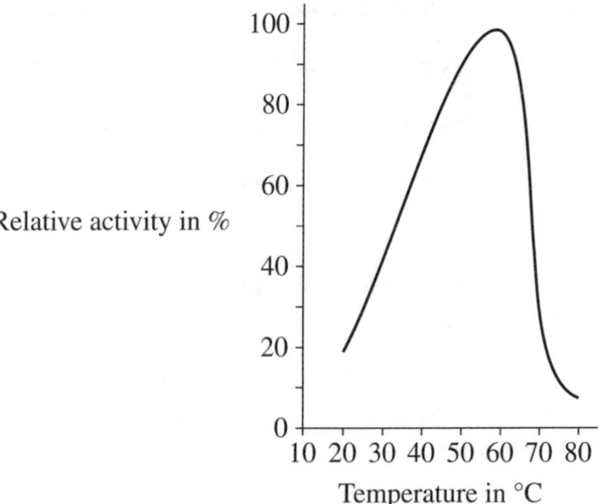

Explain the shape of the graph. **(4 marks)**

7 A patient's blood has been infected by a microorganism. The diagram shows a sample of the patient's blood seen through a light microscope.

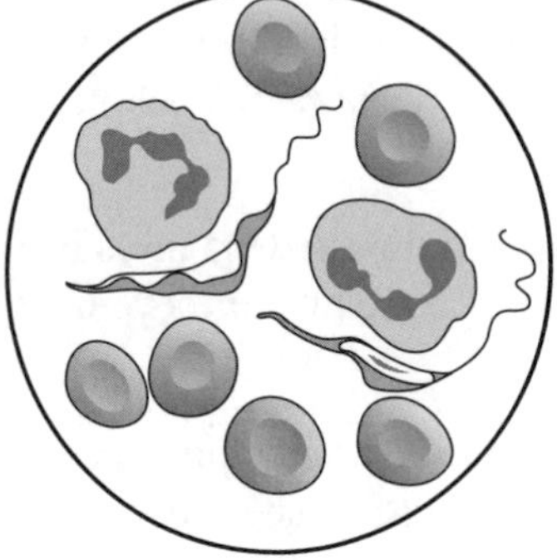

(a) How many blood cells are shown in the diagram? Tick **one** box.

2 ☐
6 ☐
8 ☐
10 ☐ **(1 mark)**

(b) Name the liquid component of blood and describe its function. **(2 marks)**

(c) Describe the function of red blood cells. **(1 mark)**

(d) Another component of blood is not shown in the diagram. It consists of fragments of cells. Name this component and describe its function. **(2 marks)**

(e) White blood cells defend the body against pathogens that have infected it. Describe **two** ways in which they do this. **(4 marks)**

8 A study was carried out to investigate whether medical drugs called statins lower the risk of cardiovascular disease. A double-blind trial was carried out using 18 000 people. Half were given statin tablets and half were given placebo tablets. The graph shows the results of the study.

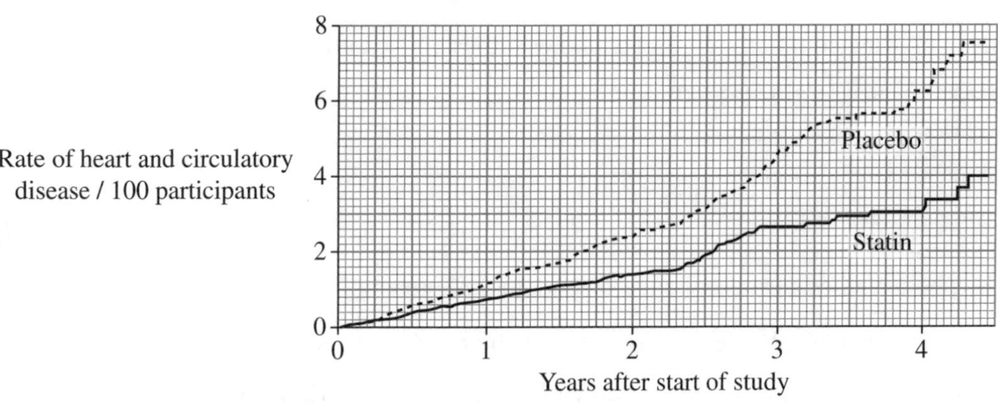

(a) Explain why a placebo is used in drug trials. **(2 marks)**

(b) Explain how reliable the results of this trial are. **(2 marks)**

(c) A doctor looked at the results from the study. The doctor decided that prescribing statins would help her patients cut their risk of heart disease. Justify this decision. **(2 marks)**

9 Plants have developed transport mechanisms to obtain and transport water, sugars and ions.

(a) Name the plant cells responsible for absorbing water and mineral ions from the soil. **(1 mark)**

(b) Plants use xylem vessels and phloem vessels for transport. Compare the functions of the two types of vessel. **(2 marks)**

(c) A student investigated the loss of water from the leaves of a plant. He used the apparatus shown in the diagram.

The student set up the apparatus and measured how far the bubble moved in 5 minutes. He repeated the experiment to obtain four measurements. The table shows the student's results.

Measurement number	1	2	3	4
Distance moved by the bubble in mm	59	68	21	62

bubble

Use appropriate data from the student's results to calculate the mean rate of movement by the bubble. Give your answer in mm/min. **(3 marks)**

(d) The student repeated his investigation in a darker room. Predict the effect on the rate of movement of the bubble, and explain your answer. **(3 marks)**

10* Monoclonal antibodies are produced from a single clone of cells. Monoclonal antibodies are specific to one binding site on one protein antigen. This means that they can target a specific substance or specific cells in the body.

Describe how monoclonal antibodies are produced, and give **two** examples of how they may be used. **(6 marks)**

11 It is believed that a type of tree, called the acacia, co-evolved with ants. If an elephant tries to eat the leaves, it shakes the tree and the ants swarm out of hollows they have dug in the thorns of the tree. This deters the elephants. The tree supplies food for the ants from small swellings near its leaves.

(a) Suggest the benefit for the acacia tree of having ants living on it. **(2 marks)**

(b) The acacia tree produces a substance that inhibits ants from feeding on newly opened flowers. This substance does not affect other insects. Suggest a reason why the tree inhibits ants from feeding on its newly opened flowers. **(2 marks)**

12 Plants are able to make their own food by the process of photosynthesis.

(a) Which of the following equations represents photosynthesis? Tick **one** box.

carbon dioxide + water → glucose + oxygen ☐
carbon dioxide + oxygen → glucose ☐
carbon dioxide + oxygen → glucose + water ☐
carbon dioxide + water → glucose ☐ **(1 mark)**

(b) Explain why photosynthesis is described as an endothermic reaction. **(2 marks)**

(c) Give **two** ways in which plants use the glucose made by photosynthesis. **(2 marks)**

(d) A farmer grew tomato plants in a greenhouse with artificial lighting. Explain what would happen to the rate of photosynthesis if the light intensity was gradually increased, but the levels of carbon dioxide and the temperature stayed the same. **(3 marks)** (e)* The diagram shows a cross-section through a leaf.

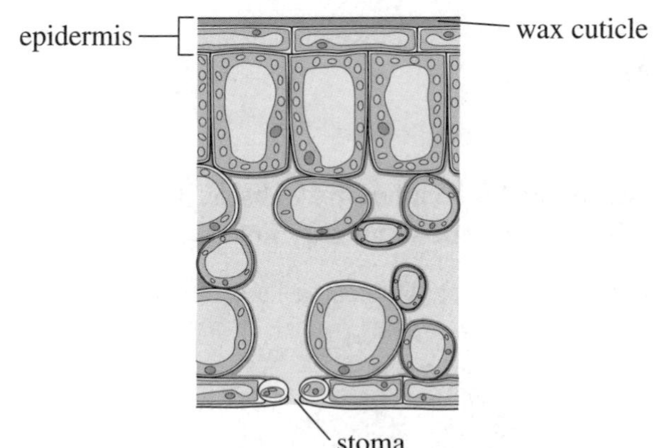

Explain how the leaf is adapted to carry out photosynthesis. **(6 marks)**

13 Fermentation is an anaerobic process that takes place in yeast cells.

(a) Which of the following equations represents fermentation? Tick **one** box.

glucose + oxygen → water + carbon dioxide ☐
glucose → lactic acid ☐
glucose → water + ethanol ☐
glucose → ethanol + carbon dioxide ☐ **(1 mark)**

(b) Give **one** use of fermentation which is economically important. **(1 mark)**

14 The graph shows how body temperature and breathing rate change during and after exercise.

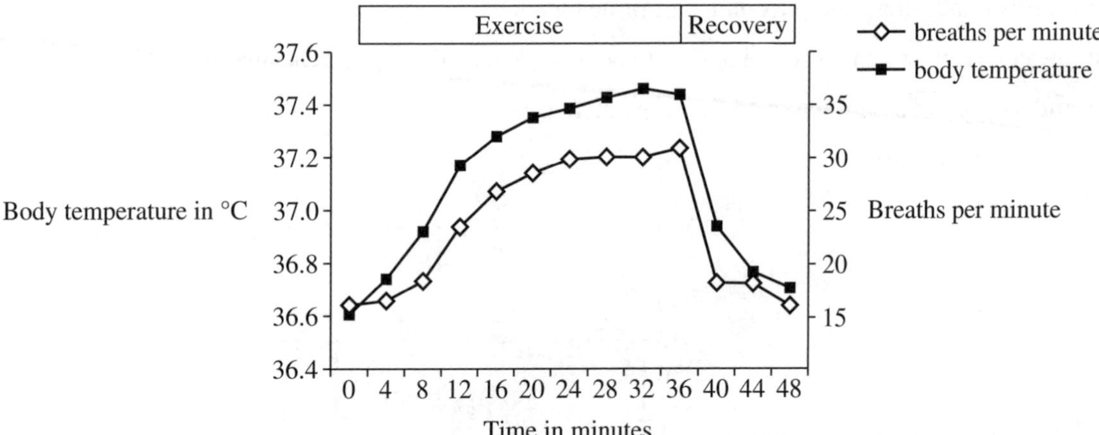

(a) Explain why the breathing rate changes during exercise. **(2 marks)**

(b) Give a reason why body temperature increases during exercise. **(1 mark)**

(c) Explain why the breathing rate does not fall to its resting value immediately after exercise. **(3 marks)**

Timed Test 2

Time allowed: 1 hour 45 minutes

Total marks: 100

AQA publishes official Sample Assessment Material on its website. This practice exam paper has been written to help you practise what you have learned and may not be representative of a real exam paper.

1 The diagram shows a hand touching a hot object. This causes a reflex action.

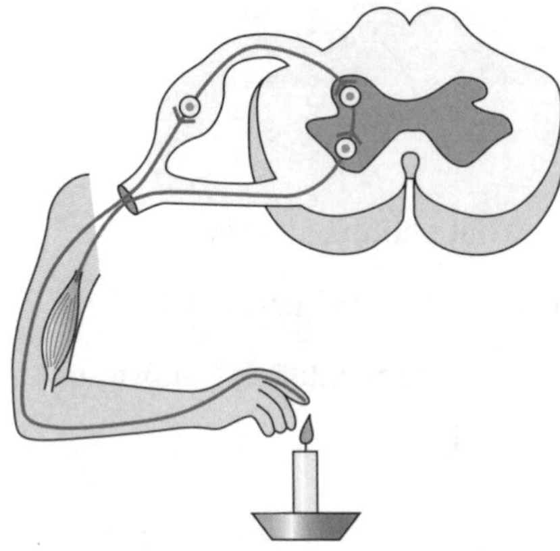

Nerve impulses result in the hand being pulled away when the hot object is touched. Describe the pathway taken by these impulses. **(3 marks)**

2 Eight students in a class carried out an experiment to investigate their reaction times. The reaction time was measured like this:

- A student is blindfolded and puts on a pair of headphones.
- A stopwatch is started and a sound is simultaneously played through the headphones.
- The student presses a button to stop the stopwatch as soon as they hear the sound.

The bar chart shows the results.

Reaction time in milliseconds (ms)

360 350 340 330 320 310 300 290 280 270 260 250

Boys Girls

(a) Calculate the mean reaction time for the boys in the class. **(2 marks)**

(b) The students' teacher says that the mean reaction time for the girls is 340 ms. Explain why the teacher uses this value for the reaction time of the girls. **(2 marks)**

(c) The class make the following conclusion: 'Boys have a faster reaction time than girls.' Evaluate their conclusion. **(3 marks)**

3 The diagram shows the main structures of the human eye.

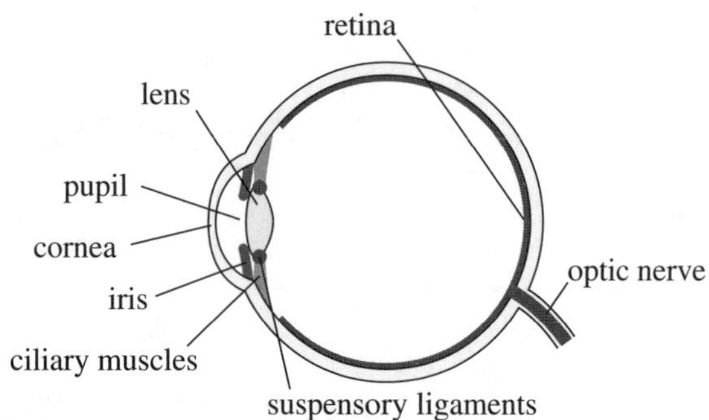

(a) Describe how the structure of the retina is related to its function in the eye. **(3 marks)**

(b) Describe how the eye accommodates to focus on a distant object. **(3 marks)**

4 Blood glucose concentration can be measured in mmol per litre. In healthy people, the concentration of glucose in the blood is maintained in the range 4–8 mmol/per litre. The concentration of glucose in the blood rises after a meal, but returns to normal levels within 2–3 hours.

(a) Describe how the body responds when the blood glucose concentration falls below 4 mmol/per litre. **(3 marks)**

(b) Glucose starts to appear in the urine if the blood glucose concentration exceeds 9 mmol/per litre. This may be a sign that a person has diabetes.

A person can be tested for diabetes. The person does not eat for 8 hours and is then given a drink containing glucose. The blood glucose concentration is then monitored for 150 minutes (2.5 hours). The graph shows the results of one of these tests.

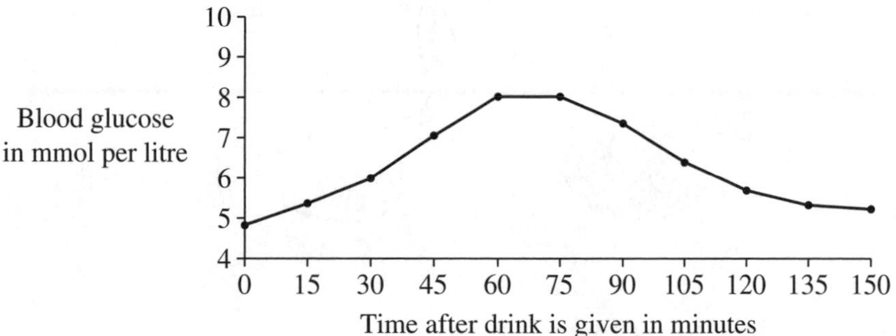

Suggest a reason why the person should not eat for 8 hours before the test. **(1 mark)**

(c) Explain whether the results shown in the graph are for a person who has diabetes. **(3 marks)**

5 During metabolism, the human body forms waste products that build up in the blood. The body is able to remove these waste products. Urea is one waste product of metabolism.

(a) Name the organ in which urea is made. **(1 mark)**

(b) Describe how urea is made. **(2 marks)**

(c) Ethanol is the alcohol in alcoholic drinks. It inhibits the release of ADH from the pituitary gland. Explain how having an alcoholic drink would affect the water content of the blood. **(3 marks)**

6 The graph shows how the levels of three different hormones vary during the menstrual cycle.

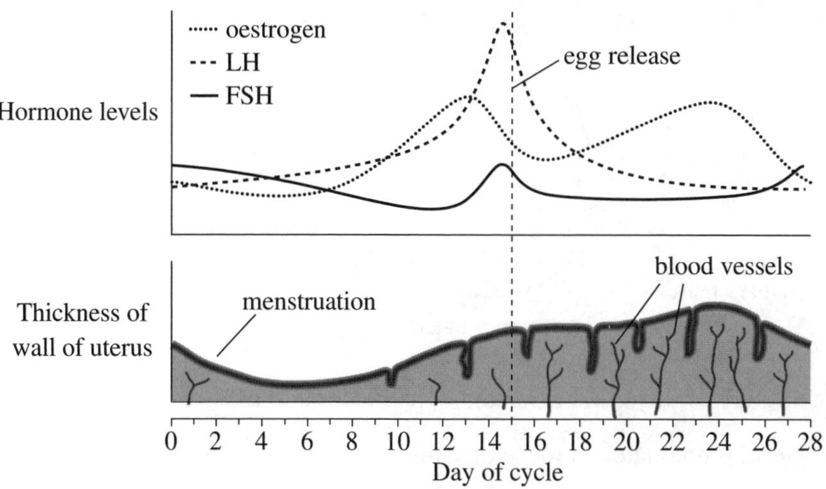

(a) Compare the effects of FSH and LH on an egg in the ovary. **(2 marks)**

(b) Describe how oestrogen interacts with the release of FSH and LH. **(2 marks)**

(c) Thyroxine is a hormone produced by the thyroid gland. It stimulates the body's basal metabolic rate. Explain how the levels of thyroxine are controlled by negative feedback. **(4 marks)**

7 Plant hormones coordinate and control the growth and responses of plants.

(a) The diagram shows the results of an experiment involving a broad bean seedling. The seedling was laid on its side at the start of the experiment (diagram **A**). Diagram **B** shows the same seedling after a few days in the dark.

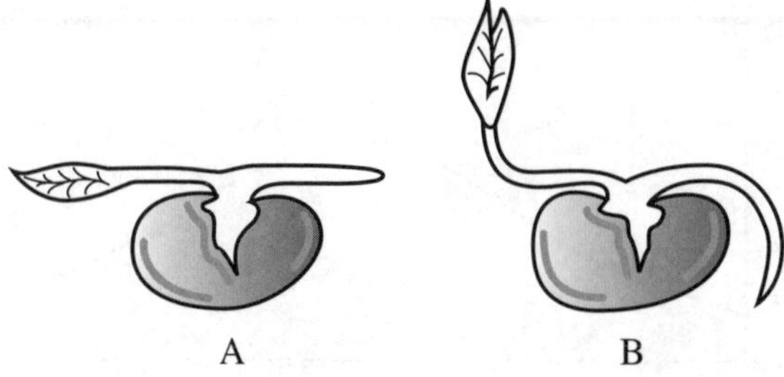

Explain the responses of the seedling seen in the diagram. **(4 marks)**

(b) Describe **two** commercial uses of gibberellins. **(2 marks)**

8 Sexual reproduction takes place when gametes from parent organisms join at fertilisation.

(a) Give the number of pairs of chromosomes in ordinary human body cells. **(1 mark)**

(b) Describe the process that results in the formation of new gametes. **(3 marks)**

(c) Draw a Punnett square diagram to determine the probability of a man and woman producing a female child. **(4 marks)**

9 Gaucher's disease is an inherited disorder which develops in early childhood. It is caused by a recessive allele. A couple want to have a baby. The couple have undergone genetic testing. Both are heterozygous for the allele.

Draw a Punnett square diagram to determine the probability of their child having Gaucher's disease. Identify any children who would have the disease, and any who would not be a carrier. Use G to represent the dominant allele and g to represent the recessive allele. **(5 marks)**

10 The genetic material in the nucleus of a cell is composed of a substance called DNA.

(a) Describe the structure of DNA. **(4 marks)**

(b) An allele contains the DNA sequence: ATCGGT. Give the complementary sequence. **(1 mark)**

(c) A polypeptide consists of 330 amino acids. Calculate the number of bases needed to code for these amino acids. **(1 mark)**

(d) Describe how a protein is synthesised using DNA. **(4 marks)**

11 The diagram describes how Lamarck explained the evolution of the long neck in the giraffe.

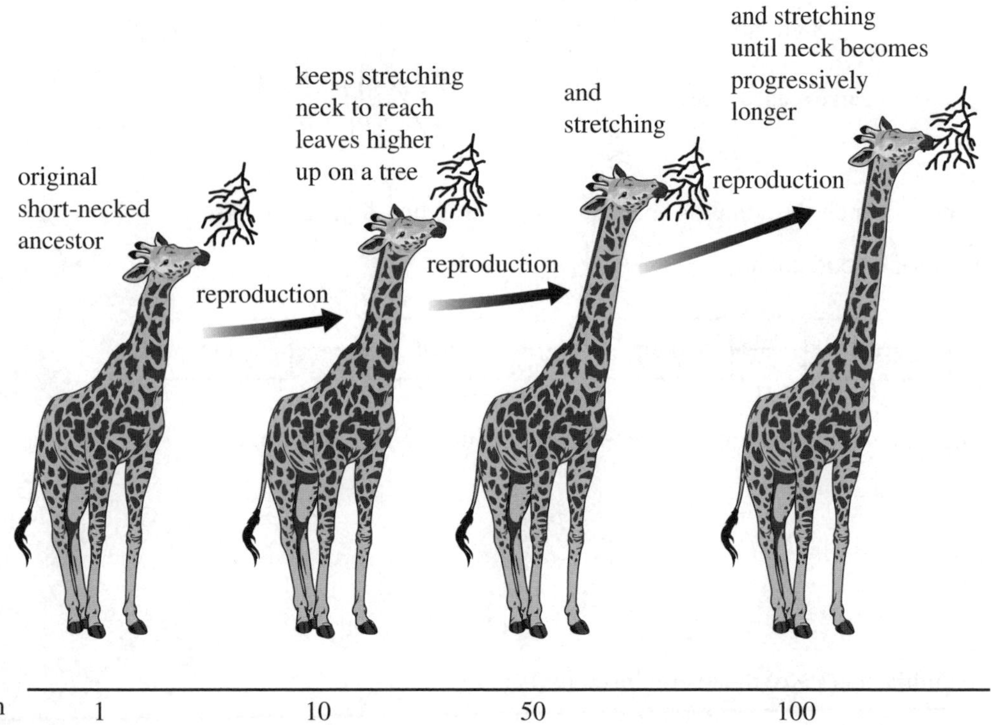

(a) Describe how Darwin would have explained the evolution of the long neck using his 'theory of evolution by natural selection'. **(4 marks)**

(b) Describe **two** reasons why Darwin's theory was only gradually accepted. **(2 marks)**

12 Lemurs are small animals that first evolved in Africa, and began to cross into Madagascar about 60 million years ago. At this time, Madagascar had just separated from the African mainland (shown in the diagram). By the time monkeys evolved and appeared in Africa, around 20 million years ago, the gap between Africa and Madagascar was too large for them to cross. In Africa, the presence of monkeys drove the lemurs towards extinction. Lemurs are not now found in the wild in Africa, but are found in Madagascar.

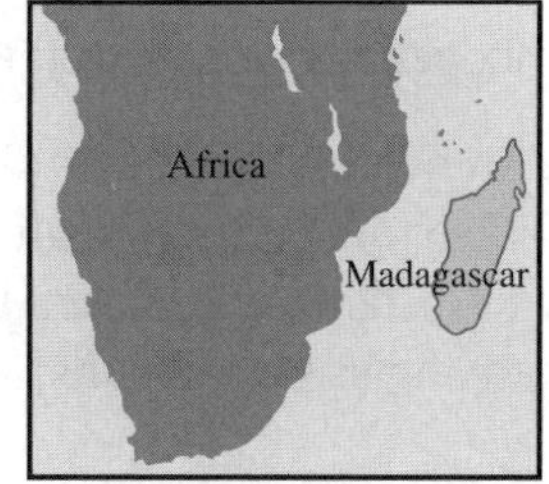

(a) Explain whether lemurs can be described as extinct. **(2 marks)**

(b) Describe how monkeys might have driven lemurs towards extinction in Africa. **(2 marks)**

(c) Scientists rely on fossils to work out how lemurs and other primates evolve. Give **one** problem that scientists may experience when using fossils for this process. **(1 mark)**

Had a go ☐ **Nearly there** ☐ **Nailed it!** ☐

13 The diagram shows a camel.

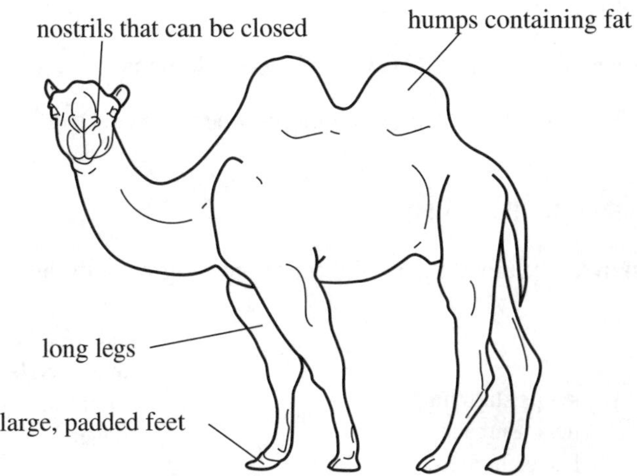

Explain **three** ways in which the camel is adapted to survive in hot, dry, sandy deserts. **(3 marks)**

14 The diagram shows a food chain.

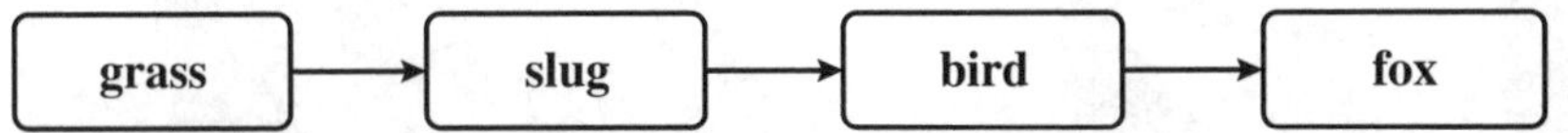

(a) Which organism in the diagram is a secondary consumer? Tick **one** box.

grass ☐
slug ☐
bird ☐
fox ☐

(1 mark)

(b) Name the initial source of the energy used by the grass. **(1 mark)**

(c) Only about 10% of the energy in one organism is passed to the next organism in the food chain. Explain why not all the energy is passed on. **(2 marks)**

(d) Sketch a pyramid of biomass for this food chain. **(2 marks)**

(e) Hedgehogs were introduced into this habitat. Hedgehogs eat slugs. Their skin is covered with spines, so they are rarely preyed on by foxes. Explain what happened to the numbers of foxes when hedgehogs were introduced. **(3 marks)**

15 A group of students noticed that different plant species were found at different distances from a main road. They did an investigation to look at how the distribution of two of these plant species changed with distance away from the road. They looked at plantain and white deadnettle plants. The bar chart shows their results.

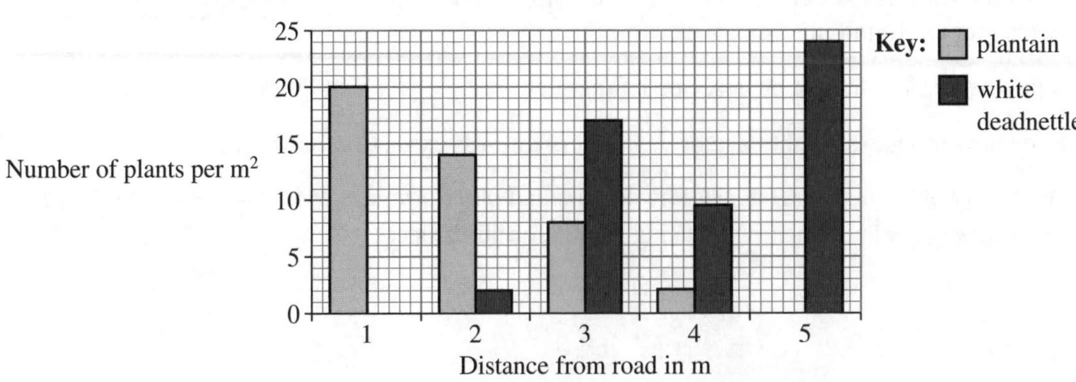

Describe how the students could have collected valid and repeatable data for this graph. **(5 marks)**

16* Killary Harbour is a 16 km fjord on the west coast of Ireland. Two types of aquaculture are carried out there, salmon farming and mussel farming.

Mussel farming involves suspending ropes in the water and mussel 'seed' (immature shellfish) are attached. The mussels then grow naturally over a period of 12–15 months. Mussels feed on microscopic organisms that they filter from the seawater. This can remove phosphates and nitrates from the water.

Some people are opposed to salmon farming because they believe it is harmful to the environment. In contrast with salmon farming, mussel farming is thought to have less impact on the environment.

Evaluate the two types of aquaculture in terms of their environmental impact. **(6 marks)**

ANSWERS

Extended response questions

Answers to 6-mark questions are indicated with a star (*).

In your exam, your answers to 6-mark questions will be marked on how well you present and organise your response, not just on the scientific content. Your responses should contain most or all of the points given in the answers below, but you should also make sure that you show how the points link to each other, and structure your response in a clear and logical way.

1. Microscopes and magnification

1 ×50 (1)

2 *The magnification of a light microscope is* usually lower / less / smaller (1) *than the magnification of an electron microscope. The level of detail seen with a light microscope is* less *than that with an electron microscope because its resolution is* less / lower / smaller (1).

3 (a) 2 μm (1)

(b) 10 μm (1)

4 (a) (1.5 × 1000)/3 (1) = ×500 (1)

(b) 3 × 750 = 2250 μm (2.25 mm) (1)

5 It lets scientists see more subcellular structures (than with a light microscope) (1); see more detail of these structures (1).

2. Animal and plant cells

1 ribosome (1)

2 (a) plant / alga (1)

(b) X: (permanent) vacuole (1)

Y: chloroplast (1)

(c) contains chromosomes / genes / DNA (1) which control the cell / control the cell's activities (1)

3 *The cell membrane controls* what enters and leaves the cell (1). *However, the cell wall is made of cellulose which* strengthens / supports / protects the cell (1).

4 contain chlorophyll / absorb sunlight (1) for photosynthesis (1) to make glucose / food for the cell (1)

3. Eukaryotes and prokaryotes

1 **1 mark** for each correct row, to **4 marks** maximum:

	Animal cells	Bacterial cells
Cytoplasm	✓	✓
Cell membrane	✓	✓
Cell wall		✓
Nucleus	✓	

2 *The chromosomal DNA is arranged to form a* loop / single loop / circle. (1) *Some bacterial cells also contain* plasmids (1).

3 200 nm; 50 μm; 1 cm; 100 mm (1)

4 (a) 2.2×10^{-3} m (1)

(b) 4.5×10^{-4} m (1)

(c) 9.7×10^{-5} m (1)

5 (a) $(2.5 \times 10^{-5})/(2.0 \times 10^{-7})$ (1) = 125 (1)

(b) –7 (1)

(c) 2 (1)

4. Specialised animal cells

1 acrosome – releases enzymes to aid entry to an egg cell (1)

nucleus – carries genetic information (1)

mitochondrion – releases energy for the cell (1)

tail – allows cell to move (1)

2 allows more space for haemoglobin / allows cell to contain more haemoglobin (1)

3 Axon is long (1) to connect to distant parts of the body (1); many / finger-like dendrites (1) make connections with other nerve cells (1).

4 *Most types of animal cells differentiate at an* early (1) *stage. As a cell differentiates, it acquires different* features / structures / subcellular structures (1) *that allow it to* carry out its function (1). *Cell division in mature animals is mainly restricted to* repair / replacement (1).

5. Specialised plant cells

1 (a) small volume cytoplasm with no nucleus (1)

(b) sieve plate with holes (1)

(c) companion cell with many mitochondria (1)

2 (a) *Xylem tissue consists of hollow tubes formed by dead xylem cells.* (1) *There are no end walls so that* water and mineral ions can flow easily through the xylem / tissue / plant (1).

(b) (i) It provides strength/support / it withstands the pressure of water moving through the tissue. (1)

(ii) Insoluble substances do not dissolve in water (1); if the lignin were to dissolve, it would not support the xylem tissue any longer (1).

3 (Most types of) animal cells differentiate at an early stage (1); (many types of) plant cells retain the ability to differentiate throughout life (1).

6. Using a light microscope

1 (a) to pass light through the slide/specimen (1)

(b) to hold the slide/specimen in place (1)

(c) to move the objective lens up or down a long way (1)

2 *The drawing is made with a pen rather than with* a pencil. (1)

any two from: the magnification is not given (1); label lines are not drawn with a ruler (1); label lines cross each other (1); cells are not drawn to scale (1); shading should not be used (except to distinguish between different structures) (1); the cell membrane cannot be seen with the light microscope (so should not be labelled) (1); outlines cross each other / are not closed for some features (1)

3 smaller field of view (1); more difficult to focus (1)

4 any three from: go back to using the low power objective (1); find the part you need and bring it back to the centre view (1); focus on it with the coarse focus (1); return to the high power objective (1) and use the fine focus wheel to bring the part into focus (1)

7. Aseptic techniques

1 16000 (1)

2 (a) to kill bacteria on the bench / so bacteria on the bench do not contaminate the culture (1)

(b) to prevent growth of anaerobic bacteria (1)

(c) Human body temperature is 37 °C so a lower temperature reduces the risk of harmful / pathogenic bacteria growing. (1)

(d) to reduce the risk of bacteria being transferred to somewhere else (1)

3 Heating kills bacteria already present in the gel (1); the gel is cooled to reduce the risk of harmful bacteria being present in the culture / bacteria cannot grow at high temperatures (1); sterilised Petri dishes prevent other bacteria contaminating the culture (1).

4 (a) *This will* keep out / stop entry of (1) *bacteria from the air that are* unwanted / not part of the experiment / likely to contaminate the culture (1).

(b) This sterilises the loop (1) to prevent existing / contaminating bacteria being cultured as well as the desired bacteria (1).

8. Investigating microbial cultures

1 (a) Bacteria do not grow in the clear zones (**1**) because the antibiotic kills them / stops them growing (**1**).

(b) Table completed, **1 mark** for each correct row:

Antibiotic	Diameter of clear zone in mm	Area of clear zone in mm^2
1	7	38.5
2	11	95.0
3	12	113
4	10	78.5

(c) antibiotic 3 (**1**), because it killed the most bacteria / prevented the most bacteria growing / formed the largest clear zone (**1**)

2 (a) *Use different concentrations* of the antibiotic (**1**); repeat with several identical plates (**1**).

(b) any two from: size of discs (**1**); volume of antibiotic solution (**1**); length of incubation time (**1**); strain of bacteria (**1**); incubation temperature (**1**)

9. Mitosis

1 two genetically identical diploid daughter cells (**1**)

2 *to produce new individuals by* asexual *reproduction* (**1**); for growth (**1**); for repair (**1**)

3 (a) A set of chromosomes is being pulled to each end of the cell (**1**); the nucleus is dividing (**1**).

(b) The cytoplasm divides (**1**); the cell membrane divides (**1**).

4 A / A and B, interphase (**1**); B, DNA replication / chromosomes are copied (**1**); C, mitosis (**1**); D, cell division (**1**)

10. Stem cells

1 an undifferentiated cell (**1**)

2 (a) meristem (**1**)

(b) (i) tip of shoots (**1**)

(ii) any two from the following for **1 mark** each: to form root hair cells (**1**); xylem (**1**); phloem (**1**)

(c) *Rare species can be cloned so they* are saved from extinction. (**1**)

Large numbers of identical plants with desirable features (e.g. disease resistance) can be produced (e.g. for farmers). (**1**)

3 (a) to become specialised (**1**); so they can carry out a particular function (**1**)

(b) any two from the following for **1 mark** each: the body may destroy / reject the donor cells (**1**); a viral infection may be transferred (**1**); stem cells may not stop dividing / risk of cancer (**1**)

(c) one advantage for **1 mark**, e.g. are not rejected by the patient's body / cells are easily extracted / cells can differentiate into more types of cells than adult stem cells

one disadvantage for **1 mark**, e.g. requires destruction of an embryo / some people have ethical/religious objections

11. Diffusion

1 *Diffusion is the* spreading out / movement (**1**) *of particles, so that there is a net movement of particles from an area of* higher concentration *to an area of* lower concentration (**1**).

2 (a) difference in concentration / concentration gradient (**1**); surface area of the membrane (**1**); thickness of the membrane (**1**)

(b) Particles move about more quickly at higher temperatures (**1**) so the rate of diffusion increases as the temperature increases (**1**).

3 The concentration of urea must be greater in the cells / lower in the plasma (**1**) because urea diffuses from the cells / into the plasma (**1**) and diffusion happens from higher to lower concentration / down a concentration gradient (**1**).

4 Substance A diffuses / net movement from the left-hand side to the right-hand side (**1**) because the concentration of its particles is greater on the left / there are five particles on the left but only two particles on the right (in the same volume) (**1**).

Substance B does not diffuse / no net movement of particles (**1**) because the concentration of its particles is same on both sides / 10 particles on both sides (in the same volume) (**1**).

12. Exchange surfaces

1 one 20-mm cube (**1**)

2 (a) villi (**1**)

(b) *Their shape gives them a large* surface area (**1**). *They provide a short diffusion path because* their surface is a single layer of cells / is thin (**1**). *A network of blood capillaries inside them ensures that* food molecules are carried away / a high concentration gradient is maintained (**1**).

3 Surface area = $6 \times 50 \times 50 = 15\,000\ \mu m^2$; volume = $50 \times 50 \times 50 = 125\,000\ \mu m^3$ (**1**)

Surface area to volume ratio = 15 000/125 000 = 0.12 *or* 3:25 (**1**)

4 The flatworm is flat and thin / the earthworm is cylindrical (**1**); the flatworm has a larger surface area to volume ratio than the earthworm (**1**); every cell in the flatworm is close to the surface (**1**); in the earthworm diffusion must happen over too great a distance / through too many layers of cells (so it needs a transport system) (**1**).

13. Osmosis

1 *Osmosis is the diffusion of* water (**1**) *from a* dilute / less concentrated (**1**) *solution to a* (more) concentrated (**1**) *solution through a* partially permeable membrane / semi-permeable membrane (**1**).

2 (a) The size at the start is a factor that affects the size at the end (**1**); it is controlled so that the procedure is valid / a fair test (**1**).

(b) Potato in distilled water: larger / longer (**1**)

Potato in strong glucose solution: smaller / shorter (**1**)

3 (a) allows small molecules (e.g. water) to pass through (**1**) but not larger molecules / ions (**1**)

(b) There is a net movement of water into the tubing (**1**) by osmosis (**1**) so the volume of liquid inside increases (**1**).

(c) The cells will expand / burst (**1**) because water moves into them by osmosis (**1**) because the water is more dilute than the cytoplasm / the cytoplasm is more concentrated than the water (**1**); and their membrane is partially permeable / semi-permeable (**1**).

14. Investigating osmosis

1 (a) missing change in mass: 0.25 (**1**); missing percentage change = $(-0.15/2.58) \times 100 = -5.8$ (**1**)

(b) x-axis scale 0 to 0.8 using the full width of the line, labelled with column heading from table (**1**); points plotted ± half square (**1**); line of best fit (**1**)

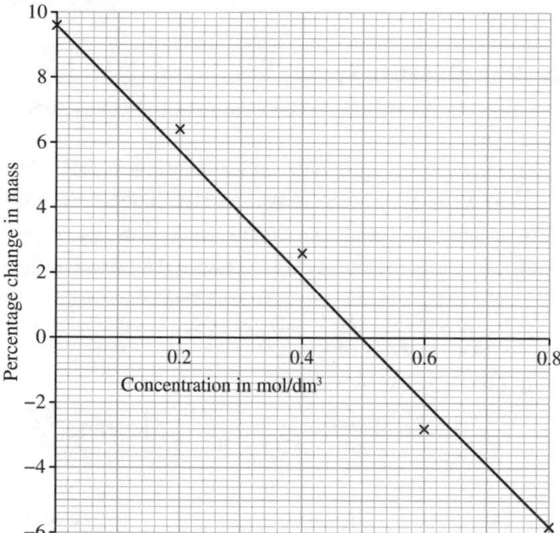

(c) answer in the range 0.45 mol/dm^3 to 0.55 mol/dm^3 (**1**)

(d) A balance with a resolution of ±0.1 g will give readings of 2.5 g and 2.7 g (**1**) so the recorded change in mass would be 0.2 g (**1**), giving a different percentage change in mass / 8% / less accurate value (**1**).

15. Active transport

1 active transport (**1**)

2 (a) Osmosis is the movement of water (not ions). (**1**)

(b) Diffusion is the net movement of particles from a higher concentration to a lower concentration (**1**); from the information given, nitrate ions are moved in the opposite direction / from lower to higher concentration (**1**).

(c) *The nitrate ions are being moved by* active transport (**1**). *This process requires* energy (**1**) *from* respiration (**1**).

3 (a) Glucose can be absorbed into the blood even if its concentration is lower in the gut / higher in the plasma. (**1**)

(b) Absorption of glucose will stop / slow down (**1**) because energy is needed for active transport (**1**) and the toxins will stop energy being released (**1**).

4 (a) one similarity, e.g. no energy (from respiration) is needed / net movement from a high concentration of particles to a lower concentration of particles (**1**); one difference, e.g. osmosis is the net movement of water only / osmosis needs a partially permeable membrane but diffusion does not (**1**)

(b) Both involve the movement of particles (**1**); two differences, e.g. diffusion is net movement from a high concentration of particles to a lower concentration, but active transport is net movement from a low concentration to a higher concentration (**1**); diffusion does not need energy from respiration but active transport does (**1**).

16. Extended response – Cell biology

*Answer could include the following points to **6 marks**:

Similarities: both have:

- cell membrane, which controls what enters and leaves the cell
- cell wall, which strengthens the cell
- cytoplasm, in which cell reactions happen
- ribosomes, where protein synthesis happens.

Differences:

- Plant cells have mitochondria, where respiration happens, but bacterial cells do not.
- Plant cells have chloroplasts, where photosynthesis happens, but bacterial cells do not.
- DNA, which contains genetic information, is contained in a nucleus in plant cells, but is found in a single chromosome / loop and in plasmids in bacterial cells.
- Cell wall in plant cells is made from cellulose but from different substances in bacterial cells.
- Plant cells have a permanent vacuole filled with cell sap, which helps to keep the cell rigid, but bacterial cells do not.

17. The digestive system

1 A, liver (**1**); B, stomach (**1**); C, pancreas (**1**); D, large intestine (**1**); E, small intestine (**1**)

2 (a) *A tissue is a group of cells with a similar* structure (**1**) *and* function (**1**) (structure and function can be given in the opposite order).

(b) aggregations / combinations of tissues (**1**) with specific functions (**1**)

(c) two from the following: nervous system (**1**); circulatory system (**1**); respiratory system (**1**); excretory system (**1**); endocrine system (**1**); reproductive system (**1**); immune system (**1**); muscular / skeletal system (**1**); exocrine system (**1**)

3 (a) **1 mark** for each correct box:

Type of enzyme	Substrate	Product(s)
carbohydrase	carbohydrates (**1**)	simple sugars
protease	proteins	amino acids (**1**)
lipase	lipids / fats (**1**)	fatty acids *and* glycerol (**1**)

(b) starch (**1**)

18. Food testing

1 (a) pestle and mortar (**1**)

(b) a red-stained layer floating on a layer of water (**1**)

2 *Put some egg white in a test tube. Add an equal volume of* Biuret reagent / solution (**1**) *and shake to mix. If protein is present, the mixture turns* pink / purple / mauve (**1**).

3 (a) Heat water in a kettle (**1**); pour hot water into a beaker and place the test tubes in the water (**1**).

(b) Both food samples contain reducing sugars (**1**) but the one that gives the red colour contains more / a higher concentration (**1**).

4 (a) Add iodine solution (**1**); turns black / blue–black if starch is present (**1**).

(b) one from: it can stain clothes / skin (**1**); it is harmful if it is breathed in / if it comes into contact with the skin (**1**)

19. Enzymes

1 *The active site in proteases matches the shape of* proteins (**1**) *but* the active site in lipases does not match the shape of proteins / is the wrong shape / fits fats / lipids instead (**1**).

2 (a) answer in the range 60–65 °C (**1**)

(b) As the temperature increases, the rate of collisions increases (**1**) between the substrate and the active site (**1**).

(c) The activity decreases (**1**) because at high temperatures the shape of the active site changes (so that the substrate can no longer fit into it) (**1**) and the enzyme is denatured (**1**).

3 Pepsin and trypsin digest proteins / amylase digests starch (**1**); the optimum pH of amylase is about 7 so it works in the mouth and small intestine (**1**); the optimum pH of pepsin is about 2 so it works in the stomach (**1**); the optimum pH of trypsin is about 8 so it works in the small intestine (**1**).

20. Investigating enzymes

1 (a) Values for rate calculated:

pH	6	8	10
Time	3.6	1.2	8.3
Rate	0.28	0.83	0.12

1 mark for all 3 correct values, **1 mark** for 2 significant figures

(b) Rate on vertical axis, scale 0.1 per cm; pH on horizontal axis, scale 1 pH unit per cm (**1**); both axes labelled using table row headings (**1**); all points plotted ±½ square (**1**); curve of best fit through all points (**1**)

(c) two from the following: use a thermostatically controlled water bath (**1**); repeat at each pH and calculate mean values (**1**); repeat at pH values close to pH 8 (**1**); record times in seconds (**1**)

21. The blood

1 plasma – carries other blood components (**1**); platelet – involved in forming blood clots (**1**); red blood cell – carries oxygen (**1**); white blood cell – part of the body's immune system (**1**)

2 (a) nucleus (**1**)

(b) haemoglobin (**1**)

(c) *Their biconcave shape gives them a large* surface area (**1**) *for diffusion to happen efficiently. They are also flexible, which lets them* fit through narrow blood vessels / capillaries (**1**).

3 urea (**1**); carbon dioxide (**1**)

4 Phagocyte: engulfs and destroys pathogens (**1**)

Lymphocyte: produces antibodies (**1**) which bind to pathogens (**1**)

5 two from: platelets respond to a wound by triggering clotting (**1**); platelets are trapped in a meshwork of fibrin / protein (**1**); clot prevents pathogens from entering (**1**)

22. Blood vessels

1 vein (**1**)

2 (a) Blood can flow easily at low pressure / a lot of blood can flow / blood can flow with less resistance (**1**).

(b) to withstand high blood pressure / to stretch as blood flows through (**1**); to regain shape afterwards (**1**), which smooths the flow of blood (**1**)

3 *The capillaries are about as wide as one red blood cell, so the distance oxygen must travel to the capillary wall is* small / short (**1**). *The walls are only one cell thick, so* the diffusion distance is small / short (**1**).

4 (a) two from: muscles (in the body) contract and press on veins (**1**); veins have valves (**1**); to keep blood flowing in the same direction / prevent back-flow (**1**)

(b) Veins have a thinner muscle wall than arteries (**1**); so it is easier to get the needle in (**1**).

OR Veins contain blood under lower pressure (**1**), so taking blood is more controlled (**1**).

23. The heart

1 **1 mark** for each correct row:

Blood vessel	Carries blood from	Carries blood to	Carries oxygenated blood (✓ or ✗)
aorta	heart	body	✓
pulmonary artery	heart	lungs	✗
pulmonary vein	lungs	heart	✓
vena cava	body	heart	✗

2 (a) It acts as a pump (**1**); muscles contract to pump the blood (**1**).

(b) vena cava → right atrium → right ventricle → pulmonary artery (**1**), lungs (**1**), pulmonary vein → left atrium → left ventricle → aorta (**1**)

3 (a) (heart valve) closes when ventricle relaxes (**1**); prevents backflow (**1**)

(b) right ventricle (**1**); pumps blood to the lungs / pulmonary artery (**1**)

(c) has to pump harder / produce more pressure (**1**) to get blood all round body (**1**), not just to lungs (**1**)

24. The lungs

1 A, trachea (**1**); B, bronchus / bronchi (**1**)

2 (a) diffusion (**1**)

(b) *There is a net movement of carbon dioxide from* the blood / plasma *to* air in the alveoli / lungs (**1**), *and a net movement of oxygen from* the air in the alveoli / lungs to the red blood cells (**1**).

(c) two from the following:

- Millions of alveoli create a large surface area (**1**) for the diffusion of gases (**1**).
- Each alveolus is closely associated with a network of capillaries (**1**) to minimise diffusion distance (**1**).
- Wall of alveolus is one cell thick (**1**) to minimise diffusion distance (**1**).

3 symptom, e.g. difficulty breathing / shortness of breath / increased breathing rate / reduced ability for physical activity (**1**); explanation, e.g. smaller surface area for gas exchange (**1**) so less oxygen absorbed (**1**)

25. Cardiovascular disease

1 (a) The coronary arteries supply blood to the heart muscle (**1**); the fatty deposits reduce the blood flow in these arteries (**1**), which reduces the amount of oxygen available to the heart muscle (**1**).

(b) (i) A stent is a wire frame inserted into the artery (**1**); it is expanded and allows blood to flow more easily (**1**).

(ii) Statins are drugs / medicines that reduce blood cholesterol levels (**1**); this slows down the rate at which fatty material is deposited (**1**).

2 (a) may not open fully (**1**); may develop a leak / not close fully (**1**)

(b) biological / human / animal valves (**1**); mechanical / artificial / metal / polymer valves (**1**)

(c) tiredness / breathlessness / death (**1**)

3 **1 mark** for each box, e.g. lifestyle changes benefit: no side effects, may reduce risk of other health problems (**1**); lifestyle changes drawback: may take a long time to work, may not work effectively (**1**); medication benefit: easier than changing lifestyle / starts working immediately / cheaper and less risky than surgery (**1**); medication drawback: may have side effects / needs to be taken long term / may interfere with other medication (**1**); surgery benefit: usually a long-term solution (**1**); surgery drawback: risk of infection / risk of complications / expensive / more difficult than medication (**1**)

26. Health and disease

1 diet / stress / life situation (**1**)

2 (a) communicable disease (**1**)

(b) (i) example of a communicable disease, e.g. colds / flu / measles / food poisoning (**1**)

(ii) example of a non-communicable disease, e.g. cardiovascular disease / cancer / diabetes (**1**)

(c) **1 mark** for each correct row:

	Communicable disease	Non-communicable disease
Number of cases	rapid variation over time	changes gradually over time
Distribution of cases	often localised	may be widespread

3 (a) Scabies is caused by a mite / arthropod (**1**), which can be passed from person to person (**1**).

(b) The itchy skin is an allergic immune response (**1**), which cannot be passed from person to person (**1**).

(c) Scratching the itchy skin breaks the skin's surface (**1**), so bacteria are more likely to enter the skin (**1**).

27. Lifestyle and disease

1 Eat less / take more exercise (**1**); because he is obese (**1**); and needs to lose mass (**1**).

2 (a) (Scientists can show that) it is a cause of the disease (**1**); know / understand how this works (**1**)

(b) cells that have changed (**1**), so they have uncontrolled growth / division (**1**)

3 (a) The risk of diabetes increases as a person's mass increases (**1**); there is a large / largest increase in risk going from overweight to obese (**1**).

(b) (i) Total percentage of people with diabetes = 0.7 + 0.9 + 1.2 + 2.1 = 4.9% (**1**)

Number of people with diabetes = 4.9 ÷ 100 × 19 500 000 (**1**) = 955 500 (**1**) (allow 0.955 million / 0.96 million / 960 000)

(ii) one from the following, for **1 mark**: the number of people studied / sampled may be small (**1**); the percentages may be rounded values (**1**); the people in the study may not represent the whole city (**1**); there may be categories of body type that are not in the chart (**1**)

28. Alcohol and smoking

1 (a) blood concentration = 5 × 20 = 100 mg per 100 cm^3 of blood (**1**); (from the graph) the increased risk is ×6 (**1**)

(b) *As the concentration of alcohol in the blood rises, the risk of having a car accident* increases / rises. (**1**) *The risk of having an accident increases as the* concentration of alcohol in the blood rises. (**1**)

(c) Alcohol causes slower reactions / blurred vision / increased risk taking. (**1**)

2 (a) total number of deaths = 123 800 + 86 800 + 90 600 + 84 600 + 17 400 + 34 700 = 437 900; percentage of deaths from lung cancer = (123 800/437 900) × 100 (**1**) = 28.3% (**1**)

(b) blood carries less oxygen to the baby (**1**); smaller birth weight (**1**)

29. The leaf

1 organ (**1**)

2 anchors the plant in the soil (**1**); absorbs water / ions from the soil water (**1**)

3 (a) (i) guard cells (**1**)

(ii) stoma / stomata (**1**)

(b) (i) *This is so that light can* pass through easily. (**1**)

(ii) one from: reduces water loss (**1**); protects the leaf (e.g. from water or microorganisms entering it) (**1**); helps to keep the upper surface clean (**1**)

(c) (i) contain a lot of chloroplasts for photosynthesis (**1**); cylindrical shape so they pack closely together / there are few cell walls for light to pass through (**1**)

(ii) Cells are packed loosely / irregularly together to produce air spaces (**1**); with large surface areas for gas exchange (**1**).

30. Transpiration

1 Mineral ions enter by active transport and water by osmosis. (**1**)

2 *Water* evaporates (**1**) *from the leaves, mostly through the* stomata (**1**). *This causes a pull so that water moves through the* plant / xylem (**1**) *and is replaced by water entering the roots.*

3 (a) Lower surface: (22 + 18 + 23 + 19)/4 = 20.5 (**1**)

Upper surface: (0 + 2 + 3 + 1)/4 = 1.5 (**1**)

(b) the lower surface because it has the most / more stomata (**1**), and most water is lost through stomata (**1**)

(c) (i) It will reduce water loss / reduce wilting. (**1**)

(ii) two from: carbon dioxide for photosynthesis is needed (**1**); oxygen for respiration is needed (**1**); gas exchange is needed (**1**); minerals are transported in the xylem by transpiration when the stomata are open (**1**); transpiration / evaporation keeps the leaf cool in hot weather (**1**)

31. Investigating transpiration

1 (a) 3.2 cm (**1**)

(b) (i) radius of tube = 0.5/2 = 0.25 mm

volume travelled = $\pi \times 0.25^2 \times 90 = 17.7$ mm^3 (**1**)

rate of transpiration = 17.7/5 = 3.5 mm^3/min (**1**) (to 1 decimal place)

(ii) rate of transpiration was higher when the fan was on (**1**); because the movement of air removes water vapour more quickly from the leaves (**1**); increasing the concentration gradient from leaf to air (**1**)

2 (a) Rate of photosynthesis is greater (**1**); stomata are more open for gas exchange (**1**).

(b) Rate of photosynthesis is greater (**1**); stomata are more open for gas exchange (**1**); water molecules move more quickly / rate of diffusion is greater (**1**).

32. Translocation

1 phloem (**1**)

2 movement of dissolved food / sucrose molecules (**1**) from leaves / storage organs to the rest of the plant (**1**)

3 **1 mark** for each correct row:

Structure or mechanism	Transpiration	Translocation
xylem	✓	
phloem		✓
pulled by evaporation from the leaf	✓	
energy needed		✓

4 (a) distance = 0.40 × 1000 = 400 mm *and* time = 67 × 60 = 4020 s (**1**); rate = 400/4020 = 0.1 mm/s (**1**)

(b) Carbon dioxide is used for photosynthesis (**1**); radioactive carbon atoms form part of glucose molecules (**1**); some glucose is converted into sucrose (**1**).

5 can differentiate (**1**), to form specialised cells / any type of plant cell (**1**)

33. Extended response – Organisation

*Answer could include the following points to **6 marks**:

Outline of route:

- right atrium → right ventricle
- right ventricle → pulmonary artery
- pulmonary artery → (capillaries in) lungs
- (capillaries in) lungs → pulmonary vein
- pulmonary vein → left atrium
- left atrium → left ventricle
- left ventricle → aorta
- aorta → rest of the body / capillaries in the body
- rest of the body / capillaries in the body → vena cava
- vena cava → right atrium

Answer might also include:

- Valves in heart / veins prevent backflow of blood.
- Deoxygenated blood enters / leaves right side.
- Oxygenated blood enters / leaves left side.
- Walls of left side of heart are thicker than right side.

34. Viral diseases

1 influenza (**1**)

2 (a) **1 mark** for each correct row:

Feature	Incorrect (✗)
Viruses are about the same size as cells.	✗
Viruses can infect plants or animals.	
Viruses reproduce outside cells.	✗
Viruses are spread by direct contact, air or water.	

(b) Viruses are much smaller than cells (**1**); viruses reproduce inside cells (**1**).

3 (a) *Chloroplasts contain* chlorophyll. (**1**) *Lighter-coloured areas appear where* there is less chlorophyll / fewer chloroplasts. (**1**)

(b) less photosynthesis (**1**), so less glucose made (**1**), and less respiration / release of energy / synthesis of other substances (**1**)

4 (a) red skin rash (**1**)

(b) Measles can be fatal if complications occur. (**1**)

(c) inhaling droplets from sneezes / coughs (**1**)

5 HIV attacks the body's immune cells (**1**), so the immune system can no longer deal with other infections (**1**).

35. Bacterial diseases

1 cholera (**1**)

2 two from: fever (**1**); abdominal cramps (**1**); vomiting (**1**); diarrhoea (**1**)

3 (a) one from: bacteria more easily enter the body during surgery (**1**); contaminated hands (**1**); contaminated equipment (**1**)

(b) Fewer patients died (**1**); one from: reduced from 45 per 100 to 15 per 100 (**1**); reduced from 45% to 15% (**1**); reduced by two-thirds (**1**); reduced to a third (**1**); reduced by 67% (**1**)

4 *Bacteria get into the body, where they* reproduce rapidly (**1**). *The bacteria produce* toxins *which damage cells / tissues* (**1**).

5 (a) one from: thick yellow / green discharge from vagina / penis (**1**); pain on urinating (**1**)

(b) Penicillin is an antibiotic (**1**); antibiotics kill bacteria / prevent bacteria dividing (**1**); resistant strains of bacteria have appeared (**1**).

36. Fungal and protist diseases

1 fungus, protist (**1**)

2 (a) fungicide (**1**)

(b) Fungus / fungal spores from an infected person could get on the towel (**1**), and pass to you when you use the towel (**1**).

3 *If the leaves are damaged or there are fewer leaves,* photosynthesis (**1**) *is reduced. The plant cannot make enough* glucose / food for healthy growth (**1**).

4 (a) one from: recurring fever (**1**); nausea (**1**); anaemia (**1**)

(b) It is a vector (**1**). It carries the pathogen / *Plasmodium* from person to person (**1**).

(c) Mosquito net prevents mosquitoes biting / reaching people (**1**), so mosquitoes cannot transfer the pathogen / *Plasmodium* (**1**) from an infected person to an uninfected person (**1**).

37. Human defence systems

1 (a) (i) Hairs trap bacteria / pathogens / harmful particles. (**1**)

(ii) Hydrochloric acid kills bacteria / pathogens in food / drink. (**1**)

(b) two from: physical barrier to pathogens (**1**); secretes antimicrobial substances / substances that kill bacteria / substances that inhibit bacterial growth (**1**); scabs form over damaged skin (**1**)

(c) (i) lysozyme (**1**)

(ii) kills bacteria (**1**); by digesting their cell walls (**1**)

2 (a) mucus (**1**)

(b) It is sticky (**1**), so it traps bacteria / pathogens (**1**).

(c) (i) cilia / cilium (**1**)

(ii) *The structures on the surface of cells move* in waves (**1**), *which move* mucus / substance A out of the lungs (**1**) towards the throat where it is swallowed (**1**).

38. The immune system

1 (a) *Antibodies are* proteins. (**1**) *They attach to* antigens (**1**) *produced by the pathogen, which leads to its destruction.*

(b) Cells / phagocytes ingest pathogens (phagocytosis) (**1**) and destroy / digest the pathogens (**1**). White blood cells produce antitoxins (**1**) which attach to toxins / poisonous substances and make them inactive (**1**).

2 (a) White blood cells that produce antibodies against the measles virus are activated (**1**); these cells divide many times (**1**), so the concentration of antibodies increases (**1**), and then decreases when the viruses have been destroyed (**1**).

(b) Some of the antibody-producing white blood cells stay in the blood as memory lymphocytes / cells (**1**); these respond / divide after infection (**1**), so the number of white blood cells producing antibodies against the measles virus increases rapidly (**1**).

(c) (The child had not been exposed in the past because) line **B** is similar in size and shape to line **A** (**1**), which was for a first infection with measles / the line would be higher if it was a second infection (**1**).

(d) The concentration of antibodies increased faster / to a higher concentration (**1**), so the measles viruses were destroyed before they could cause illness / symptoms / disease (**1**).

39. Vaccination

1 (a) (small amounts of) dead or inactive forms of the pathogen (**1**)

(b) side effect, e.g. soreness / swelling / mild symptoms of disease (**1**); may give only partial protection if there are different strains of the pathogen (**1**)

(c) *The vaccine causes white blood cells to make* antibodies (**1**) *against the pathogen. If the same pathogen enters the body again, white blood cells respond* quickly / rapidly (**1**) to produce the correct antibodies / antibodies against the pathogen to prevent illness (**1**).

2 (a) 2003 (**1**)

(b) The number of cases would increase (**1**), because fewer babies were immunised which would have protected them from infection (**1**).

(c) The pathogen can be transmitted only from an infected individual (**1**); vaccinated people are immune and so cannot become infected (**1**); if most people are immune, the chances of an unvaccinated person meeting an infected person are very small (**1**).

40. Antibiotics and painkillers

1 (a) Antibiotics kill bacteria / prevent bacteria reproducing (**1**), without damaging body cells / tissues (**1**).

(b) *The pharmacist's advice would be* not to take (**1**) *penicillin. This is because antibiotics* do not kill / are not effective against *viruses* (**1**).

2 (a) Number of deaths increased between 2000 and 2005 (then levelled off) (**1**), then declined after 2006 (**1**).

(b) new antibiotics invented (**1**); to which the MRSA bacteria are not resistant (**1**)

OR improved hygiene in hospitals (**1**) has helped prevent spread of MRSA (**1**)

(c) Bacteria mutate / variation in the population (**1**); some bacterial cells are resistant to the antibiotic (**1**); resistant bacteria survive to reproduce (**1**).

3 Viruses exist / reproduce inside cells (**1**); drugs that kill viruses may also harm the body's tissues (**1**).

41. New medicines

1 aspirin (**1**)

2 (a) 3, 1, 5, 2, 4 (all correct = **2 marks**, 3 or 4 correct = **1 mark**)

(b) two from: check that substance is not toxic (**1**); check its efficacy / how well it works at treating the disease (**1**); determine the dose / how much is needed for the drug to work (**1**)

(c) testing in cells / tissues (to see if it has a desired effect) (**1**); testing on animals (to see how it works in a whole body / has no harmful side effects) (**1**)

3 (a) dummy drug / looks like the drug but does not contain any of it (**1**)

(b) (It is effective because) it appears to be effective in nearly 400 people with high blood pressure (**1**); the reduction is much greater than those in the placebo group / it has very little adverse effect on the blood pressure of those in the 'normal' group (**1**).

42. Monoclonal antibodies

1 (a) All the antibodies are identical (**1**); Large amounts of antibody are produced (**1**).

(b) *Cells called* lymphocytes (**1**) *are collected from mice. The cells are stimulated to make a particular* antibody (**1**). *These cells are then fused with a* tumour / myeloma (**1**) *cell to make a* hybridoma (**1**) *cell.*

2 The antibody is specific for a particular substance / antigen. (**1**)

3 (a) Monoclonal antibodies are made that attach to antigens present only on cancer cells. (**1**)

Radioactive substance is attached to the antibodies (**1**); the location of the antibodies is detected using a scanner / radiation detector (**1**).

OR (Fluorescent) dye is attached to the antibodies (**1**); the location of the antibodies is detected (in a tissue sample / biopsy) using a microscope (**1**).

(b) A toxic / radioactive substance is attached to a monoclonal antibody (**1**) that only binds to / recognises antigens on cancer cells (**1**), so the substance is delivered to cancer cells (**1**) but not to healthy cells that do not have the antigen (**1**).

43. Plant disease

1 (a) fungus (**1**)

(b) insect (**1**)

2 **1 mark** for each correct box:

	Needed for	Symptom of deficiency
Magnesium ions	making chlorophyll	chlorosis (loss of green colour)
Nitrate ions	making proteins	stunted growth

3 (a) one from: stunted growth (1); areas of decay / rot (1); abnormal growths (1); poorly formed stems / leaves (1); presence of pests (1)
(b) He could have looked at a website / gardening manual. (1)
(c) Different diseases may cause the same symptoms. (1)
(d) (i) so she could analyse the soil (1) to look for soil factors / nutrient deficiency / pH / toxins (1)
(ii) so she could test for the presence of pathogens (1) using test kits containing monoclonal antibodies / by microscopy / genetic analysis / culturing the pathogen (1)

44. Plant defences

1 *The cell membranes are surrounded by* cellulose cell walls (1). *The epidermal layer is covered by a* tough waxy cuticle (1).

2 (a) two from: time (1); temperature (1); volume of garlic–water mixture and water (1); species of bacteria (1); number of bacteria at the start (1); volume of bacterial culture (1)
(b) to act as a control (1); to check that any change was due to the garlic (1)
(c) to defend against **bacterial** pathogens (1)

3 (a) When an insect lands on the leaf / walks on the leaf (1), the leaf curls / droops and the insect falls off (1).
(b) one from: thorns / hairs to deter animals (1); mimicry to trick animals (1)

45. Extended response – Infection and response

*Answer could include the following points to **6 marks**:

Differences:

- Vaccines stimulate the body's immune system but antibiotics kill bacteria / inhibit their cell processes / stop them growing.
- Vaccines must be given before infection but antibiotics can be given after infection with a pathogen.
- Vaccinated people become immune to a disease without getting it but the pathogen must already be present for antibiotics to work.
- Vaccines can protect against viruses as well as bacteria but antibiotics are effective only against bacteria.
- Herd immunity means not everyone in a population needs to be vaccinated but all infected people need to be given an antibiotic against a particular pathogen.
- Vaccines prevent reinfection by the same pathogen but antibiotics do not.

Similarities:

- Some vaccines and antibiotics may give only partial protection (e.g. if there are different strains of the pathogen).
- Some people develop side effects, e.g. soreness and mild symptoms of disease with a vaccine, diarrhoea with antibiotics.
- Vaccines may fail to protect against different strains of a pathogen.

46. Photosynthesis

1 an endothermic process in which energy is transferred to the chloroplasts by light (1)

2 carbon dioxide + water → glucose + oxygen (1)

3 (a) starch / fat / oil / lipid (1)
(b) cellulose (1) to strengthen the cell wall (1)
OR amino acids (1) to make proteins (1)

4 (a) light intensity (1); amount / mass of chlorophyll (1)
(b) For each graph, correct axis labels (1); correct shape of line (1) (to a maximum of **4 marks**):

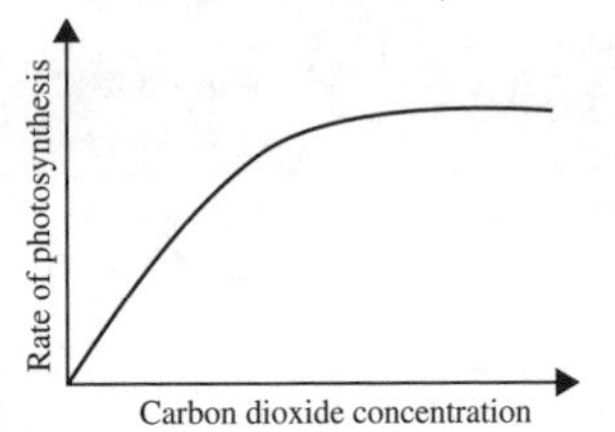

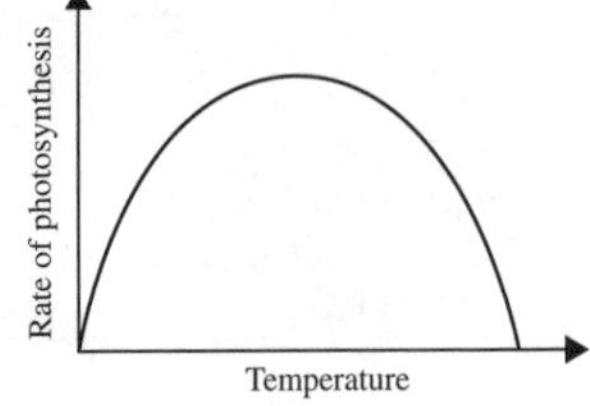

47. Limiting factors

1 *This is a factor or variable that stops the rate of something* increasing / changing (1). *The rate will increase only if this factor is* increased (1).

2 (a) Increasing the carbon dioxide concentration increases the rate of photosynthesis (1); rate of photosynthesis is proportional to carbon dioxide concentration (over the range used) (1).
(b) Increase the temperature. (1)
(c) Carbon dioxide is a limiting factor (1); increasing carbon dioxide concentration increases the rate of photosynthesis (1) so the yield will increase (1).

3 Increased temperature increases the rate of reaction (1) so photosynthesis / growth happens faster (1); eventually other factors limit rate / rate reaches maximum (1); higher temperatures denature enzymes responsible for photosynthesis (1).

48. Investigating photosynthesis

1 (a) (i) Values for 1/distance calculated and entered into the table (1).
(ii) Values for 1/(distance)^2 and entered into the table (1).

Distance, d, in m	0.30	0.25	0.20	0.15	0.10
$\frac{1}{\text{distance}}$	3.3	4	5	6.7	10
$\frac{1}{(\text{distance})^2}$	11	16	25	44	100
Number of bubbles	14	32	54	72	88

(b) graph plotted: suitable axis scales and labelled (1); 5 points correctly plotted (2) *but* 4 points correctly plotted (1); line of best fit (1)

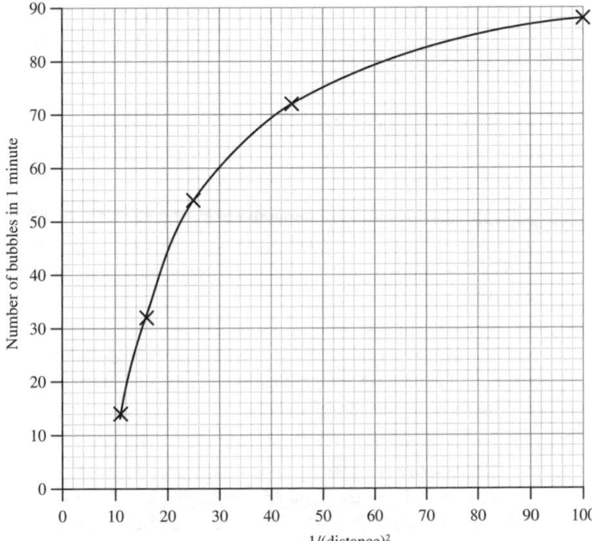

(c) Photosynthesis produces oxygen (1); the greater the rate of photosynthesis, the greater the rate of oxygen production (1); and the bubbles contain oxygen (1).
(d) the greater the light intensity, the higher the rate (1); not a linear relationship (1)

49. Respiration

1 glucose + oxygen → carbon dioxide + water (1)

2 (a) *keeping warm* (1); two from: movement (1); chemical reactions to build larger molecules (1); active transport (1)
(b) It is an exothermic reaction because energy is transferred to the surroundings (1), and energy is transferred from the surroundings in an endothermic reaction (1).

3 two from: plants cannot use energy from sunlight directly for metabolic processes (1) so they need energy from respiration for this purpose (1) during the day as well as at night (1)

4 (a) fermentation (1)
(b) three from: both happen in the absence of oxygen (1); both use glucose / both release small amounts of energy (compared with aerobic respiration) (1); yeast produces carbon dioxide but muscles do not (1); yeast produces ethanol but muscles produce lactic acid (1)

(c) four from: both use glucose / release energy (**1**); aerobic respiration produces carbon dioxide / water but anaerobic respiration does not (**1**); anaerobic respiration produces lactic acid but aerobic respiration does not (**1**); aerobic respiration releases more energy than anaerobic respiration (**1**); aerobic respiration needs oxygen but anaerobic respiration does not (**1**)

50. Responding to exercise

1 (a) 100 – 80 = 20 beats/min (**1**)

percentage increase = (20/80) × 100 = 25% (**1**)

(b) Respiring cells need oxygen (**1**); increased exercise increases the demand for oxygen (**1**); pulse / heart rate increases to supply the muscles with more oxygen / oxygenated blood (**1**).

2 (a) three from: oxygen consumption increases during exercise (**1**) but reaches a maximum value (**1**); no more oxygen can be delivered for aerobic respiration (**1**); increased energy needed comes from anaerobic respiration (**1**)

(b) three from: lactic acid builds up (**1**); after exercise, extra oxygen is needed to break down lactic acid (**1**); this is the oxygen debt / excess post-exercise oxygen consumption / EPOC (**1**), so oxygen consumption does not fall rapidly after exercise (**1**)

51. Metabolism

1 the sum of all the reactions in a cell or the body. (**1**)

2 1 *lactic acid (produced during anaerobic respiration) into* glucose (**1**)

2 *excess glucose into* glycogen (**1**) *for storage in the liver*

excess amino acids into ammonia *(then into* urea) (**1**)

3 (a) nitrate (**1**)

(b) protein / polypeptide (**1**)

(c) starch (**1**); cellulose (**1**)

4 (a) glycerol (**1**); fatty acids (**1**)

(b) one glycerol molecule *and* three fatty acid molecules (**1**)

5 carbohydrase / amylase – carbohydrates / starch (**1**); lipase – lipids / fats / oils (**1**); protease – proteins (**1**)

52. Extended response – Bioenergetics

*Answer could include any of the following points to **6 marks**:

Similarities – both processes:

- take place in cells / sub-cellular structures
- involve the transfer of energy
- involve chemical reactions
- involve glucose / oxygen / carbon dioxide / water
- require diffusion / gas exchange to obtain / release gases.

Differences:

- photosynthesis: carbon dioxide + water → glucose + oxygen / $6CO_2 + 6H_2O \rightarrow C_6H_{12}O_6 + 6O_2$
- aerobic respiration: glucose + oxygen → carbon dioxide + water / $C_6H_{12}O_6 + 6O_2 \rightarrow 6CO_2 + 6H_2O$
- Carbon dioxide and water are the reactants for photosynthesis, but the products of aerobic respiration.
- Glucose and oxygen are the products of photosynthesis, but the reactants for aerobic respiration.
- Photosynthesis is endothermic / takes in energy from the surroundings, but aerobic respiration is exothermic / gives out energy to the surroundings.
- Energy is transferred as light for photosynthesis, but (for example) by heating for aerobic respiration.
- Photosynthesis takes place in chloroplasts, but aerobic respiration takes place in mitochondria.

Paper 2

53. Homeostasis

1 *Homeostasis is the regulation of the* internal conditions (**1**) *of a cell or organism to maintain* optimum conditions (**1**) *in response to* internal and external (**1**) *changes.*

2 (a) detect stimuli (**1**), which are changes in the environment (**1**)

(b) one from: brain (**1**); spinal cord (**1**)

(c) (i) muscles (**1**); glands (**1**)

(ii) bring about responses (**1**), which restore optimum levels (**1**)

3 (a) blood glucose concentration (**1**); water level (**1**)

(b) The rate of reaction depends upon temperature (**1**); reactions would be slow / slower / too slow at low temperatures (**1**).

(c) Enzymes become denatured at high temperatures (**1**), so they do not work (**1**).

54. Neurones

1 (a) carry electrical / nerve impulses from one part of the CNS to another / carry impulses from sensory neurones to motor neurones (**1**)

(b) carry electrical / nerve impulses from the CNS to effectors / muscles / glands (**1**)

2 *The axon and dendron are long so the neurone can* transmit impulses over long distances (**1**); there are several axon terminals that can pass impulses to other neurones (**1**); dendrites collect impulses from receptor cells (**1**); myelin sheath insulates the neurone so the impulse cannot cross from adjacent neurones / impulse speed is increased (**1**).

3 (a) Myelin sheath speeds up transmission (**1**) because the impulse jumps from one gap to another (**1**).

(b) Their movement would be impaired / made difficult (**1**) because the nerve impulses would be slower / nerve impulses may pass to adjacent neurones (**1**).

55. Reflex actions

1 (a) synapse (**1**)

(b) neurone **Y**, because it carries impulses to an effector / muscle (**1**)

(c) *Neurone X releases a* neurotransmitter (**1**) *which* diffuses across the synapse / gap (**1**), *causing* neurone **Y** to generate an electrical impulse (**1**).

2 in any order: innate (**1**); rapid (**1**); automatic (**1**)

3 (a) a change in the environment (**1**)

(b) In order: stimulus is detected by receptors (**1**); nerve impulse travels through a sensory neurone (**1**); through a relay neurone (in the central nervous system) (**1**); then through a motor neurone to the effector (which produces a response) (**1**).

56. Investigating reaction times

1 (a) mean = (193 + 186 + 190 + 184 + 181 + 176)/6

= 1110/6 = 185 mm (**1**)

(b) (i) 176 mm (**1**) because this was the shortest distance / the ruler fell for the shortest time (**1**)

(ii) distance = 193/10 = 19.3 cm

reaction time = $\sqrt{19.3/491}$ (**1**)

= 0.198 (**1**) = 0.20 s (**1**) (to 2 significant figures)

(c) The drop distance decreased between the first and last drop (**1**), showing that the drop time decreased (**1**).

57. The brain

1 (a) **X**, **Y**, **Z** (second row) (**1**)

(b) (i) regulates heartbeat / breathing / unconscious processes (**1**)

(ii) coordinates and controls precise and smooth movement (**1**)

(c) two from: controls voluntary movement (**1**); interprets sensory information (**1**); responsible for learning (**1**); responsible for memory (**1**)

2 (a) *The scientists can electrically* stimulate (**1**) *different parts of the brain. They can also study people with* brain damage (**1**) *and use* MRI techniques (**1**).

(b) one from: the brain is very complex (**1**) so it is difficult to be sure of the function of each region (**1**); the brain is very delicate (**1**) so it is easily damaged (**1**)

58. The eye

1 A – lens (**1**); B – retina (**1**); C – optic nerve (**1**); D – *ciliary muscles* (**1**); E – iris (**1**)

2 three from: retina contains light-sensitive receptor cells (**1**); rods are sensitive to low light (**1**); cones are sensitive to bright light / colours (**1**); sends nerve impulses to the optic nerve (**1**)

3 (a) Light passes through the cornea (**1**), the pupil (in the iris) (**1**), then the lens (to the retina) (**1**).

(b) Ciliary muscles contract (**1**); suspensory ligaments loosen (**1**); lens becomes thicker (**1**).

4 The pupil size is increased (**1**); it lets more light through (**1**), so the patient may be dazzled in bright light (**1**).

59. Eye defects

1 **1 mark** for each correct row:

Statement	Myopia	Hyperopia
Near objects appear clear but distant objects appear blurred.	✓	
The eyeball is shorter than it should be.		✓
The lens cannot become curved enough to correct the problem.		✓

2 (a) two from: contact lenses (**1**); laser eye surgery (**1**); replacement lens in the eye (**1**)

(b) long-sightedness / hyperopia because the lens is convex / converging (**1**)

(c) *The lens refracts the light before it enters the eye, so that it* converges (**1**) so that the light focuses onto the retina (**1**).

(d) diagram showing: rays diverging through spectacle lens (**1**); converging through the cornea and lens (**1**); meeting on the retina (**1**)

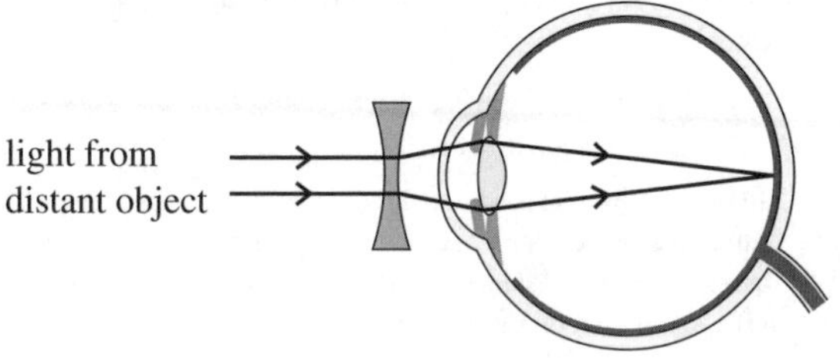

60. Thermoregulation

1 (a) Skin contains temperature receptors (**1**), which send nervous impulses to the thermoregulatory centre in the brain (**1**).

(b) Skeletal muscles repeatedly contract and relax (**1**), releasing / transferring energy by heating due to increased respiration (in the muscle cells) (**1**).

2 (a) two from: body and skin temperature are the same / 37 °C before exercise (**1**); during exercise core body temperature remains stable / does not change (**1**); during exercise skin temperature rises / varies / rises to a maximum of 38 °C (**1**)

(b) Sweat glands secrete sweat onto the skin (**1**); evaporation of water / sweat causes a transfer of energy from the skin / to the environment (**1**).

(c) *When the temperature rises, blood vessels* dilate (**1**). *There is more* blood flow through the skin capillaries (**1**). *This means that more* energy is transferred to the environment by heating (**1**).

61. Hormones

1 **A** – pituitary (**1**); **B** – thyroid (**1**); **C** – pancreas (**1**); **D** – testis / testes (**1**); **E** – ovary (**1**); **F** – adrenal (**1**)

2 **1 mark** for each correct row:

Hormone	Produced in	Target organ
ADH	pituitary gland	kidneys
adrenaline	*adrenal gland*	*heart / muscles*
FSH and LH	pituitary gland	ovaries
glucagon	pancreas	liver / muscle
oestrogen	ovary / ovaries	ovaries / uterus / pituitary gland
progesterone	ovary / ovaries	uterus
testosterone	testis / testes	male reproductive organs
TSH	pituitary gland	thyroid gland

3 Endocrine system signals are hormones carried in the blood, but nervous system signals are electrical carried in neurones (**1**); endocrine effects are slower / nervous system effects are faster (**1**); endocrine effects are longer lasting / nervous system effects are shorter lasting (**1**)

62. Blood glucose regulation

1 (a) Blood sugar level increases (**1**); an example given from the graph, e.g. blood glucose levels go from 4 mmol/dm^3 to 6 mmol/dm^3 after breakfast (**1**).

(b) Blood glucose concentration lowers more rapidly (**1**), because muscles use more glucose for respiration during exercise (**1**).

2 (a) A and F – pancreas (**1**); B – insulin (**1**); C and D – liver (**1**); E – glucagon (**1**)

(b) (i) Insulin stimulates the liver to take in excess glucose (**1**); excess glucose is converted to glycogen (**1**).

(ii) Glucagon stimulates the liver to convert glycogen back to glucose (**1**); glucose is released by the liver back into the blood (**1**).

63. Diabetes

1 (a) Total percentage with type 2 diabetes = 3.2 + 7 + 12.6 = 22.8% (**1**)

Number of people = 22.8 × 1 864 035/100 = 425 000 (**1**)

(b) one from: the sample size may be small (**1**); the sample may not accurately represent the whole population (**1**); the percentages are rounded values / not precise (**1**)

(c) As the BMI increases the percentage of people with Type 2 diabetes increases (**1**); more than half the people with Type 2 diabetes are obese (**1**).

2 (a) Type 1 diabetes: the pancreas fails to produce sufficient insulin / pancreatic cells are destroyed by the body's immune system (**1**). Type 2 diabetes: liver and muscle cells no longer respond to insulin (**1**).

(b) injections of insulin (**1**) to reduce the levels of blood glucose (**1**)

(c) carbohydrate-controlled diet (**1**); exercise (**1**)

64. Controlling water balance

1 X Y Z (first row) (**1**)

2 via the lungs during exhalation (**1**); from the skin in sweat (**1**)

3 They would gain water / expand / burst (**1**); by osmosis / water is more dilute than cytoplasm / cytoplasm is more concentrated than water (**1**).

4 (a) The kidneys help maintain the balance of water and mineral salts (**1**) so different amounts may be reabsorbed (**1**).

(b) (i) *Protein molecules are too* large (**1**) *to pass through the* pores in the kidneys / to be filtered out of the plasma (**1**).

(ii) Glucose leaves the blood in the kidneys (**1**) and is then reabsorbed (**1**).

(iii) Urea leaves the blood in the kidneys (**1**) but is not reabsorbed / most is not reabsorbed (**1**).

65. ADH and the kidneys

1 (a) digestion of proteins from the diet (**1**)

(b) (i) ammonia (**1**)

(ii) ammonia is toxic (**1**); urea is (safely) excreted (**1**)

2 (a) pituitary gland (**1**)

(b) (i) kidney tubules (**1**)

(ii) Permeability (of tubules) to water is decreased (**1**) so less water is reabsorbed (**1**) and volume of urine is increased (**1**).

3 (a) *The volume of urine would be* smaller (**1**) *and the urine would be more* concentrated (**1**).

(b) any four from: water lost from body as sweat (due to running / hot day) (**1**); so water content of blood decreased (**1**); pituitary gland secreted more ADH (**1**); more ADH in the bloodstream (**1**); more water reabsorbed from tubules back into the blood (**1**)

66. Kidney treatments

1 (a) receiving a kidney transplant from a living relative (**1**)

(b) answer in the range 14–16% (**1**)

(c) Using a machine in hospital leads to difficulties keeping a job because: appointments need to be made (**1**); these will tend to be during the working day (**1**), and will happen frequently (**1**).

OR

Using a machine at home makes it easier to keep a job because: you can use the machine in your own time (**1**), so you can avoid using it during the working day (**1**), and have the flexibility to use it when you like, e.g. overnight (**1**).

2 It separates blood from dialysis fluid (**1**); allows small molecules to diffuse into the fluid (**1**).

3 the same as (**1**); the same as (**1**); less than (**1**); more than (**1**)

67. Reproductive hormones

1 (a) testosterone (**1**)

(b) one from: increases growth / size of testes (**1**); stimulates sperm production (**1**)

(c) two from: growth spurt (**1**); underarm / pubic hair begins to grow (**1**); external genitals grow (**1**); voice deepens (**1**); brain matures (**1**)

2 (a) menstruation (**1**)

(b) (i) luteinising hormone (**1**); *stimulates the release of an egg* (**1**)

(ii) follicle-stimulating hormone (**1**); causes maturation of an egg in an ovary (**1**)

(c) ovulation (**1**)

(d) oestrogen (**1**); progesterone (**1**)

68. Control of the menstrual cycle

1 **One mark** for each correct row:

Hormone	Site of production		
	Ovaries	Pituitary gland	Empty egg follicle
oestrogen	✓		
FSH		✓	
LH		✓	
progesterone			✓

2 (a) oestrogen (**1**)

(b) (i) stimulates release because LH levels go up (**1**)

(ii) inhibits / reduces release because FSH levels go down (**1**)

(c) Release is inhibited / reduced because levels of FSH and LH go down (**1**).

(d) (It does not) because level of oestrogen falls around day 24 (**1**) and this would trigger menstruation (**1**).

69. Contraception

1 (a) vasectomy (**1**)

(b) (i) condom (**1**); diaphragm (**1**)

(ii) Spermicidal agents kill / disable sperm. (**1**)

2 (a) *FSH causes an egg to* mature (**1**) *and LH causes an egg to* be released (**1**). *A missed pill causes the level of progesterone to* decrease (**1**), *so* (FSH and LH are not inhibited and) a mature egg is released (**1**).

(b) more likely to be effective (**1**) because a daily dose cannot be missed (**1**) OR longer-lasting contraception (**1**) because progesterone is released slowly (**1**)

(c) Non-smokers on this pill are 5 times more likely (40 vs 8) to develop blood clots than non-smokers not on the pill (**1**); this is a smaller risk than is present in smokers or pregnant women (**1**). Overall level of risk is very small at only 40 in 100 000 non-smokers / This pill appears to be relatively safe to use / This pill is less safe for smokers to use (**1**).

70. Treating infertility

1 (a) *LH stimulates* ovulation / the release of an egg from the ovary. (**1**)

(b) Increase in temperature coincides with ovulation / release of an egg (**1**) so she will be more fertile in the days after (**1**).

(c) Several mature eggs may be released at once (**1**) leading to multiple births (**1**).

2 (a) to stimulate eggs to mature / to stimulate the production of oestrogen (**1**)

(b) eggs collected from the mother (**1**); eggs fertilised with sperm from the father in the laboratory (**1**); fertilised eggs develop into embryos / tiny balls of cells (**1**); one or two of these are inserted into the mother's uterus (**1**)

3 two benefits from: it allows couples to have children if they are not able to naturally (**1**); even though success rate is low, it is still successful for some couples (**1**); it can be easier to use this procedure than to adopt (**1**)

two drawbacks from: the success rate for IVF is quite low (**1**); success rate is (12 400 / 45 250) × 100 = 27% (**1**); cost is quite high (especially if more than one cycle of treatment is needed) (**1**); total cost to the NHS in 2010 = 45 250 × £2500 = £113 million (**1**)

71. Adrenaline and thyroxine

1 (a) thyroid gland (**1**)

(b) increases the basal metabolic rate (**1**)

2 (a) TSH levels increase. (**1**)

(b) An increase in thyroxine concentration causes changes (**1**) that bring about a decrease in the amount of thyroxine released (**1**) (or vice versa).

3 (a) Adrenaline is produced by the adrenal glands (**1**) in times of fear / stress (**1**).

(b) (i) increases the heart rate (**1**)

(ii) *The delivery of* oxygen *and* glucose (**1**) *to the brain and* muscles (**1**) *increases. This prepares the man's body for* 'flight or fight' (**1**).

(c) Adrenal glands stop producing adrenaline when the danger is past (**1**), so levels of adrenaline decrease (**1**), without a corrective mechanism to keep the levels within a small range (**1**).

72. Plant hormones

1 (a) positive phototropism (**1**)

(b) tip of root / shoot labelled X (**1**)

2 *High concentrations of auxins cause elongation of cells in* shoots (**1**) *but inhibit* cell elongation in roots (**1**).

3 (a) The root will grow downwards (**1**), because roots grow in the direction of the force of gravity (**1**).

(b) Plant roots grow downwards where there is more water (**1**); this helps to anchor the plant into the ground (**1**).

(c) Auxin accumulates on the lower surface of the root (**1**); it inhibits elongation of the cells here (**1**), so the upper surface of the root becomes longer, making the root bend downwards (**1**).

73. Investigating plant responses

1 (a) (102 +93 +99)/3 = 98 cm (**1**)

(b) Gibberellins cause the plant to grow taller / faster (**1**). Correct manipulation of numbers, e.g. the mean growth for plants sprayed with gibberellins, was 35 cm taller than the plants sprayed with water (**1**).

(c) It was the control (**1**); to eliminate other factors that may affect the growth of the plants / to ensure that growth was only influenced by gibberellins (**1**).

2 *Auxin is made in the tip of the shoot, so the shoot with its tip removed* does not contain auxins / contains equal concentrations of auxins on both sides / does not respond to the light (**1**); the normal shoot grows towards the light (**1**), because auxin accumulates in the dark side / the side furthest from the light source (**1**), elongating cells on that side of the shoot (**1**).

74. Uses of plant hormones

1 (a) ethene (**1**)

(b) Fruit does not become overripe when travelling long distances (**1**). It has a longer shelf-life / reaches shops in 'just-ripened' condition (**1**).

2 (a) The roots will grow more slowly than before (**1**). The shoots will grow more quickly than before (**1**), because auxins stimulate elongation in shoot cells but inhibit elongation in root cells (**1**).

(b) The broad-leaved plants grow too quickly / develop poor root structure / die (**1**), and no longer compete with the crop plants for water / minerals / light (**1**), so the crop plants grow / develop better (**1**).

3 Spray the plants with gibberellins (**1**), which cause the plants to flower early (**1**).

4 *Rooting powder contains* auxins (**1**), *which are plant hormones that cause cuttings to develop* roots (**1**). *This means that the growers can* produce clones / identical copies of the original plant (**1**).

75. Extended response – Homeostasis and response

*Answer could include some of the following points to **6 marks**:

- The thyroid gland produces thyroxine.
- Increased thyroxine levels cause reduced levels of TSH.
- Reduced levels of TSH should cause reduced levels of thyroxine.
- However, an overactive thyroid gland would cause more thyroxine to be produced.
- Thyroxine stimulates the basal metabolic rate.
- High levels of thyroxine cause high metabolic rate.
- More energy from food is needed by the body.
- Stores of body fat are used to meet the additional energy needs (so the patient loses body mass).

76. Meiosis

1 *A male gamete (a* sperm *cell)* fuses *with* a female gamete / egg (**1**) *to form a* zygote (**1**).

2 Chromosomes are copied / DNA is replicated. (**1**)

3 four daughter cells drawn (**1**); each cell with a different combination of large / small and black / white chromosomes (**1**), e.g.

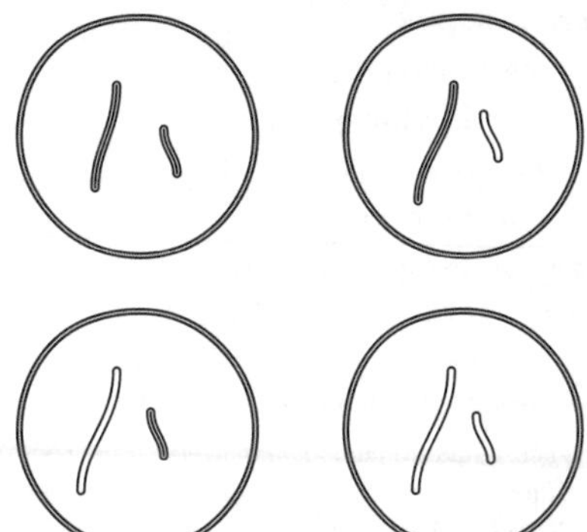

4 (a) 10 chromosomes (**1**)

(b) Each daughter cell has a copy of one chromosome from each pair of different chromosomes (**1**); different daughter cells have different combinations of these chromosomes (**1**).

77. Sexual and asexual reproduction

1 **1 mark** for each correct row:

	Sexual reproduction	Asexual reproduction
Need to find a mate	yes	no
Mixing of genetic information	mixes genetic information from each parent	no mixing of genetic information
Characteristics of offspring	show variety	same characteristics as parent / each other

2 (a) Runners: asexual (**1**) because only one parent / offspring all identical to parent (**1**).

Fruits: sexual (**1**) because there are two parents / fusion of gametes (**1**).

(b) (i) Benefit: plant can make use of beneficial conditions because more plants produced quickly. (**1**)

Drawback: no genetic variation so plants may die / not grow as well if conditions change. (**1**)

(ii) Benefit: offspring may be better adapted if conditions change because of variation / prevents overcrowding because seeds spread more widely. (**1**)

Drawback: requires two parents (pollination) / requires energy to produce fruit. (**1**)

3 two from: energy needed to find a mate (**1**); in courtship behaviour (**1**); in gamete production (**1**)

78. DNA and the genome

1 27 (**1**)

2 (a) genome (**1**)

(b) *A gene is a small section of DNA* (**1**) *which* codes for a specific protein (**1**).

3 (a) double helix (**1**)

(b) (i) A, base (**1**); B, sugar / deoxyribose (**1**); C, phosphate group (**1**)

(ii) The structure consists of repeated monomers / nucleotides. (**1**)

4 Order of bases: TACCCG (**1**)

Explanation: because there are complementary base-pairs (**1**); C always pairs with G, T with A (**1**)

79. Protein synthesis

1 (a) 2, 1, 6, *3*, 5, 4

All six correct for **3 marks**; four or five correct for **2 marks**; two or three correct for **1 mark**.

(b) folds up (**1**) to form a unique shape (**1**)

2 (a) *There can be a small change in the sequence of bases in the DNA*; (**1**) Part of the DNA may be deleted / added (**1**); Part of the DNA may be repeated (**1**).

(b) Non-coding parts of DNA can switch genes on and off (**1**), so variations in these areas of DNA may affect how genes are expressed (**1**).

3 A change in the sequence of bases (**1**) leads to a change in the amino acid sequence of the protein (**1**), so that it has a different structure from the normal protein (**1**).

80. Genetic terms

1 (a) a sex cell (**1**) produced by meiosis (**1**)

(b) different forms (**1**) of a gene (**1**)

2 *The genotype of an organism is the alleles of* a particular gene present (**1**). *However, an organism's phenotype is its* observed characteristics / traits (**1**) *produced by* these alleles (working at a molecular level) (**1**).

3 (a) (i) Bb (**1**)

(ii) (brown) because the B allele is dominant (**1**), so only one copy is needed for it to be expressed (**1**)

(b) bb (**1**); to have blue eyes she must have two recessive alleles (**1**)

4 The R allele is dominant (**1**) because it is expressed if there are two copies of the allele or only one (**1**). The r allele is recessive (**1**) because it is expressed only if there are two copies of the allele (**1**).

81. Genetic crosses

1 (a) dd (**1**); the d allele is recessive so two copies are needed for poorly formed wings (**1**)

(b) Completed diagram:

		Parent 1	
		D	d
Parent 2	D	DD	Dd
	d	Dd	dd

parent gametes correct (**1**); offspring genotypes correct (**1**); dd circled in the diagram (**1**); probability = 0.25 / 25% / ¼ / 1 in 4 / 1:3 (**1**)

2 (a) Completed diagram:

		Parent 1	
		G	g
Parent 2	g	Gg	gg
	g	Gg	gg

Parent gametes Gg (**1**) and gg (**1**) (*either order*); offspring genotypes correct (**1**)

(b) Grey fur: Gg

White fur: gg (**1**)

82. Family trees

1 Both parents are heterozygous for the sickle-cell allele. (**1**)

2 (a) two / 2 (**1**)

(b) one / 1 / individual 5 only (**1**)

(c) *Person 4 doesn't have cystic fibrosis but they must have inherited an* f (**1**) *allele from their mother. So they must have inherited an* F (**1**) *allele from their father. This means that person 4's genotype is* Ff (**1**).

(d) Two healthy parents (person 3 and person 4) (**1**) produce a child (person 8) with CF (**1**).

83. Inheritance

1 (a) 23 (**1**)

(b) completed Punnett square diagram:

		Father	
		X	Y
Mother	X	XX	XY
	X	XX	XY

parent gametes correct (**1**); offspring genotypes correct (**1**)

(c) female (**1**)

2 *The graph shows that the chance of a woman having a child with Down's syndrome* increases *as the mother's age increases* (**1**). *So pregnant mothers over the age of 40 are* more (**1**) *likely to be offered embryo screening.*

3 Punnett square diagram:

		Parent 1	
		P	p
Parent 2	P	PP	Pp
	p	Pp	pp

parent gametes correct (**1**); offspring genotypes correct (**1**); pp circled in the diagram (**1**); probability = 0.25 / 25% / ¼ / 1 in 4 / 1:3 (**1**)

84. Variation and evolution

1 **1 mark** for each correct row:

	Mutations		
	Very few	Some	Most
Have no effect on phenotype			✓
Influence the phenotype		✓	
Determine the phenotype	✓		

2 (a) environmental (**1**)

(b) environmental (**1**)

(c) combination / genetic and environmental (**1**)

3 There is variation / differences in the characteristics of individuals in a population (**1**); individuals who are better suited to their environment are more likely to survive (**1**).

4 A new species forms when two populations become too different to interbreed (**1**); this takes many generations (**1**).

5 Some individuals in the population have longer necks (**1**); these individuals can reach food higher up in the trees than the others (**1**); they are more likely to reproduce and pass on their alleles to their offspring (**1**); these offspring are more likely to have longer necks (**1**).

85. Selective breeding

1 *It is the process by which humans breed* plants / animals (**1**) *for particular* genetic characteristics / traits (**1**).

2 one from: high milk yield (**1**); high / low in milk fat (**1**); high in calcium (**1**)

3 (a) two from: high yield (**1**); high nutrient content (**1**); low fertiliser needs (**1**); resistance to pests (**1**)

(b) one from: drought resistance (**1**); tolerance to high temperature (**1**)

4 (a) one from: to obtain more light for photosynthesis (**1**); to avoid shading from other plants (**1**)

(b) one from: short-stemmed plants are less likely to fall over (**1**); greater proportion of growth goes into the seeds (**1**)

5 Select the pigs in the group that have the least body fat (**1**); breed from these pigs (**1**); choose pigs with the least body fat from the offspring and breed from them (**1**); repeat this process over many generations (**1**).

6 two from: breeds may be more likely to have disease (**1**); breeds may be more likely to have inherited / genetic defects (**1**); useful alleles may be lost from the population (**1**)

86. Genetic engineering

1 changed their genome and their phenotype (**1**)

2 *The gene from a* jellyfish (**1**) *is cut out using* enzymes (**1**). *This gene is transferred to a* mouse (**1**) *embryo cell, and inserted into a chromosome. The embryo is then allowed to develop as normal.*

3 (a) Eating golden rice can increase vitamin A production / levels (**1**), which reduces the chance of vitamin A deficiency / poor immune response / illness from infection / difficulty seeing at night (**1**).

(b) The gene that is modified might transfer from the GM crop into wild rice / non-GM strains of rice (**1**) where it could have unknown effects (**1**).

OR

Eating GM crops might have an effect on human health (**1**), although this possibility has not been researched enough / risks are not known (**1**).

4 four from: GM bacteria produce human insulin not pig insulin (**1**), so this will be more effective / is the right form of insulin / is less likely to cause adverse reactions (**1**); it can be produced in large quantities by the bacteria (**1**); this means that it can be produced at low cost (**1**); some people would have ethical / religious objections to having insulin from pigs (**1**)

87. Stages in genetic engineering

1 vector (**1**)

2 (a) (i) restriction enzyme (**1**)

(ii) DNA ligase (**1**)

(b) circle of DNA (**1**); containing the added / insulin gene (**1**)

(c) any two from: to produce matching sticky ends (**1**); to produce complementary bases (**1**); so insulin gene can be incorporated into plasmid (**1**)

(d) *If there were non-GM bacteria in the fermenter as well, they might use up* nutrients / food (**1**) *or they might* produce harmful substances / cause contamination / grow faster than the GM bacteria / out-compete the GM bacteria (**1**).

88. Cloning

1 (a) Specialised / differentiated cells cannot form different tissues (**1**), and the embryo must develop into a whole animal (**1**).

(b) The new embryos come from a single original embryo (**1**), which has a different genome / genotype / DNA sequence from the host mothers (**1**).

2 (a) genetically identical individuals (**1**)

(b) *to help to preserve rare plant species* (**1**), *and to* produce many plants in plant nurseries (**1**)

(c) A small piece of tissue is cut from the parent plant (**1**); this is placed in growth medium containing plant hormones / nutrients (**1**); the tissue develops into tiny plantlets, which are planted into compost (**1**).

3 The nucleus is removed from an unfertilised egg cell (**1**); the nucleus from a body cell from the adult dog is inserted into the empty / enucleated egg cell (**1**); an electric shock stimulates the egg cell to divide to form an embryo (**1**); the embryo is allowed to develop into a ball of cells (**1**), which is then inserted into the uterus of an adult female dog to continue its development (**1**).

89. Darwin and Lamarck

1 (a) 1865 (**1**)

(b) The number of light moths decreases over time (**1**), whereas the number of dark moths increases (**1**).

(c) 1 *Not all trees were darkened, so some light moths were still camouflaged.* (**1**)

2 one from: mutation formed the alleles for the light colour (**1**); moths from other populations flew to the study area (**1**); light-coloured moths changed their behaviour to avoid being seen (**1**)

2 (a) two from: it challenged the idea that God made all the organisms on Earth (**1**); there was insufficient evidence at the time to convince many scientists (**1**); how inheritance and variation work was not known until some time after the theory was published (**1**)

(b) two from: observations from an expedition (**1**); experiments (**1**); discussions with other scientists (**1**); use of knowledge about fossils / geology (**1**)

3 Lamarck – arms stretch through use (**1**); Darwin – variation in arm length in the ape population (**1**); Lamarck – offspring inherit longer arms (**1**); Darwin – apes with longer arms better adapted to the environment so more likely to survive / to pass on their alleles (**1**)

90. Speciation

1 Darwin and Wallace (**1**)

2 (a) *Speciation is the process in which a population changes to form varieties that cannot* interbreed / breed with other varieties (**1**) to produce fertile offspring (**1**).

(b) The land separated the porkfish into two populations (**1**); natural selection in the different seas favoured different variations of characteristics (**1**); the characteristics of the two populations changed over time (**1**).

3 Environment / conditions slightly different in each area around the valley (**1**); natural selection produced slightly different species (**1**); each neighbouring species similar enough to interbreed (to produce infertile offspring) going around the valley (**1**); however, *E. eschscholtzi* and *E. klauberi* too different to interbreed (at all) (**1**).

91. Mendel

1 (a) Mendel would know what units / factors / traits / alleles they had (**1**). If he crossed two plants with different traits, there would be no other traits contributing to the outcome (**1**).

(b) Mendel would know which plants had been crossed (**1**) because pollen from other plants would be excluded (**1**).

(c) *This allowed Mendel to collect* several sets (**1**) *of data. It helped to reduce the effects of* random errors / anomalous results (**1**).

2 (a) Percentage = 100 × 349/(349 + 112) (**1**) = 75.7% (**1**)

(b) The unit for yellow seeds was dominant to the unit for green seeds (**1**). The units are inherited unchanged (**1**), because plants producing green seeds were produced in the second generation from plants that produced yellow seeds (**1**).

3 Cross dwarf plants with tall plants (**1**); use pure-breeding plants (**1**); collect seeds from the first generation and allow them to grow (**1**); all the plants should be tall if the unit for dwarf is recessive (**1**).

92. Fossils

1 (a) 1 *low temperatures* (**1**)

2 absence of oxygen / water (**1**)

(b) remains / traces of organisms from millions of years ago (**1**), which are found in rocks (**1**)

2 Black rats carried a parasite (**1**), to which they were immune, but the native rats were not (**1**).

3 (a) three from: new disease (**1**); new competitors (**1**); new predators (**1**); changes to the environment, e.g. climate (**1**)

(b) Mammoths and humans existed at the same time (**1**). Mammoths became extinct but humans did not (**1**), so something else must have caused the extinction of mammoths (**1**).

4 The fossil record is incomplete because: early forms of life had soft bodies and left few traces (**1**); traces / fossils destroyed by geological changes (**1**); conditions not present for fossils to form (**1**); fossils of small / delicate organisms less likely to survive or be discovered (**1**).

93. Resistant bacteria

1 *No, because evolution can also produce new* strains / breeds / varieties. (**1**)

2 (a) an antibiotic (**1**), because it is used to treat bacterial infections / penicillin is an antibiotic (**1**)

(b) (i) MRSA already resistant to meticillin / other antibiotics (**1**), so no / few / fewer antibiotics are left to treat vancomycin-resistant MRSA (**1**)

(ii) Mutations produce bacteria in the population that are resistant to vancomycin (**1**); these bacteria are not killed in the presence of vancomycin (**1**); they reproduce / divide (**1**); remaining non-resistant bacteria in the population are killed (**1**).

3 Doctors should not prescribe antibiotics inappropriately (**1**), such as treating non-serious or viral infections (**1**).

OR

Patients should complete their course of antibiotics (**1**) so that all bacteria are killed / none survive to mutate and form resistant strains (**1**).

4 Evolution occurs in the presence of a change / feature in the environment (**1**); resistant strains develop in the presence of antibiotics (**1**); changes are inherited (**1**).

94. Classification

1 (a) Completed table:

Classification group	Humans	Wolf	Panther
kingdom (**1**)	Animalia	Animalia	Animalia
phylum (**1**)	Chordata	Chordata	Chordata
class	Mammalia	Mammalia	Mammalia
order	Primate	Carnivora	Carnivora
family (**1**)	Hominidae	Canidae	Felidae
genus	Homo	Canis	Panthera
species (**1**)	sapiens	lupus	pardus
binomial name	Homo sapiens	*Canis lupus* (**1**)	*Panthera pardus* (**1**)

(b) panther and wolf (**1**), because they both belong to the same order (**1**)

2 Viruses are regarded as non-living (**1**) because they do not show any of the life processes / named life process / they rely on host cells to reproduce (**1**).

3 (a) *Melogale* (**1**)

(b) All the badgers belong to a different genus / species (**1**); they are not closely related despite all being called badgers (**1**).

(c) any two from: no confusion about which organisms being referred to / no two species have the same binomial name (**1**); correct organism is easily identified from a similar species (**1**); it can help scientists to study / conserve species (**1**)

95. Evolutionary trees

1 one from: amoebae have a nucleus (**1**) but bacteria do not (**1**); bacteria have a cell wall (**1**) but amoebae do not (**1**); amoebae have mitochondria (**1**) but bacteria do not (**1**)

2 Completed table:

Domain	Type(s) of organism in domain
archaea	primitive bacteria (that usually live in extreme environments) (**1**)
bacteria	true bacteria and cyanobacteria (**1**)
eukaryota	protists (**1**), plants (**1**), animals (**1**), fungi (**1**)

3 (a) toad (**1**)

(b) mouse, rat (**1**)

(c) They share common characteristics (**1**), because they evolved separately from a recent common ancestor on a branch of the evolutionary tree (**1**).

96. Extended response – Inheritance, variation and evolution

*Answer could include of the following points to **6 marks**:

Using the graph:

- The numbers of dark-form moths decreased over time.

Explaining the change:

- The population of peppered moths contains two different forms / colours / variants / kinds.
- dark form and light form
- In 1959/mid-twentieth century most / 95% were the dark form.
- The two forms are controlled by genes / alleles
- which individual moths inherit.
- The trunks of trees where moths rest in the day are part of their environment.
- When these trunks are covered in soot the dark moths are less likely to be seen and eaten by birds.
- Dark moths are more likely to survive and reproduce than the light moths.
- When soot emissions stopped tree trunks became light coloured again, so dark moths were now more likely to be seen and eaten by birds.
- Dark moths became less likely to survive and reproduce.

97. Ecosystems

1 (a) two from: water (**1**); mineral ions (**1**); space (**1**)

(b) *territory and* (**1**) food / mates (**1**)

2 community – all the different populations in a habitat (**1**)

organism – a single living individual (**1**)

population – all the organisms of the same species in a habitat (**1**)

ecosystem – all the living organisms and non-living parts in an area (**1**)

3 (a) two from: more lichens grow on the tree than on the concrete post (**1**); more lichens grow on the sides facing the Sun than away from / side on to the Sun (**1**); the most lichens are found on the tree facing the Sun (**1**); the fewest lichens are found on the post facing away from the Sun (**1**)

(b) Lichens on the trees can obtain more nutrients from the trees (**1**) but less from the concrete (**1**). Lichens on the sides facing the Sun receive the most light (**1**) for photosynthesis (**1**).

98. Interdependence

1 *All the species and* environmental factors (**1**) *are in balance so that the population sizes* stay fairly constant (**1**).

2 (a) owl (**1**)

(b) two from: grasshoppers (**1**); mice (**1**); rabbits (**1**)

(c) The number of snakes would decrease (**1**) because snakes eat only mice / there would be less food for the snakes (**1**).

(d) The number of mice will decrease (**1**) because the numbers of stoats and snakes increase (**1**), and because stoats and snakes are not being eaten (by owls) so more mice are eaten (by stoats and snakes) (**1**).

99. Adaptation

1 (a) extremophiles (**1**)

(b) Bacteria provide food (**1**) to allow other organisms to survive there (**1**), whereas the rest of the seabed does not have organisms that produce food (e.g. plants) (**1**).

2 (a) *Roses produce flowers to attract insects. These flowers have very bright* colours / petals (**1**) *and they also give off a strong* scent / smell (**1**).

(b) They have a light, feathery structure (**1**), so can easily be blown away / distributed by the wind (**1**).

3 Animal A: any three from: A has a thick(er) coat for insulation (**1**); A has dark skin in order to absorb as much energy from the Sun as possible (**1**); A has wide feet in order not to sink through the snow / spread its mass on ice (**1**); A has small(er) ears so as to preserve heat / not radiate heat (**1**).

100. Food chains

1 (a) (i) grasshopper / mouse / rabbit (**1**)

(ii) owl (**1**)

(b) grass → mouse → snake → owl (**1**)

(c) It is the producer (**1**); it makes glucose by photosynthesis (**1**).

2 *The number of aphids rises if there is plenty of food and few ladybirds, because more survive to reproduce* (**1**). *More aphids mean more food for the* ladybirds, so more ladybirds survive to reproduce (**1**); as the number of ladybirds rises, they eat more aphids, so the number of aphids falls (**1**); there is less food for the ladybirds, so the number of ladybirds falls (**1**).

101. Fieldwork techniques

1 (a) mean = (11 + 12 + 7 + 12 + 8 + 8 + 12)/7 = 70/7 = 10 (**1**)

(b) median = 11 (**1**)

(c) Quadrats have a known area (**1**); slugs are slow moving (so easily counted in the quadrat) (**1**).

2 Set up a transect line (e.g. string / tape) between the path and the woodland (**1**); place quadrats at regular intervals along the transect (**1**); count the number of these plants in each quadrat (**1**).

3 Area of the football pitch = 100 × 65 = 6500 m^2 (**1**)

Number of clover plants = 7 × 6500 = 45 500 (**1**)

102. Field investigations

1 (a) (He is correct) because the percentage cover is greater on the south side (**1**), and this is the side with the higher light intensity (**1**).

OR

(He is not correct) because although the percentage cover is greater on the side with the higher light intensity (**1**), other factors may be responsible for the difference (**1**).

(b) one from: temperature (**1**) because it affects enzymes / rate of reactions (**1**); humidity / rainfall (**1**) because water required for photosynthesis / other cell processes (**1**)

2 (a) (13 + 8)/2 = 10.5 (**1**)

(b) *As the distance from the sea increases, the number of limpets* decreases (**1**); this decrease is linear with the distance / it decreases by 4 limpets every 0.5 m (**1**); limpets are more likely to survive if they live closer to the sea (**1**).

103. Cycling materials

1 (a) photosynthesis (**1**)

(b) respiration (**1**)

(c) combustion (**1**)

(d) decomposition (**1**)

2 *Water vapour escapes from plants by* transpiration (**1**) *and from rivers and soil by* evaporation (**1**). *If the water is not replaced by rainfall,* the rivers will dry up / there will not be enough water for plants / animals / water must be conserved by restricting water use (**1**).

3 Microorganisms are decomposers (1); complex carbon-containing molecules release carbon dioxide into the atmosphere (from respiration) (1).

4 Fish and plants respire (1); respiration releases carbon dioxide into the water (1); plants absorb carbon dioxide (1), which they use in photosynthesis (1).

104. Decomposition

1 (a) *Microorganisms that decay the garden waste need aerobic conditions* (1), *and* water / moisture (1), and warm temperatures (1).

(b) Compost contains minerals / ions (1), which can easily be released and taken up by plants (1).

2 (a) mass = 7.5 × 60/100 (1) = 4.5 million tonnes (1)

(b) (i) in the absence of oxygen / air (1)

(ii) methane (1)

(iii) fuel, e.g. for cooking / heating (1)

3 Drying: removes water (1); decomposer microorganisms need water for cellular processes (1)

Refrigeration: cold temperatures so microorganisms grow more slowly (1); enzyme activity / chemical reactions reduced at low temperatures (1)

Packing in nitrogen: replaces / removes oxygen (1); decomposer microorganisms need oxygen for respiration (1)

105. Investigating decay

1 (a) *The scientists buried the bags next to each other, so the conditions would be the same.* (1) *They also used* leaves from the same tree / a single tree. (1)

(b) Decrease in mass is due to decomposition in the leaf / leaf being eaten by organisms in the soil (1), but the loss of mass is greater in the bag with large mesh (1), because the leaves can be more easily eaten by larger organisms in the soil, e.g. earthworms, woodlice / the pieces of decayed leaves can fall out when the bag is moved (1).

2 (a) answer in the range 36–38 °C (1)

(b) The rate increases up to a maximum (at about 37 °C) (1), because the rate of collisions between the substrate and enzyme / active site increases (1); at higher temperatures rate decreases (1), because the shape of the active site changes / enzymes are denatured (1).

106. Environmental change

1 (a) 29 °C (1)

(b) difference in percentage of males = 10%; difference in number of males = 120 × 10/100 = 12 (1) fewer (1)

2 (a) *amount of chlorophyll,* temperature (1) *and* light intensity (1)

(b) four from: the mass of carbon dioxide absorbed depends on the rate of photosynthesis (1); the rate of photosynthesis increases when the temperature / light intensity increases (between May and July) (1); photosynthesis occurs when the tree has leaves (e.g. between May and October) (1); the rate of photosynthesis decreases when the temperature / light intensity decreases (between September and October) (1); photosynthesis stops when the tree sheds its leaves (e.g. from October/November) (1)

107. Waste management

1 (a) *Pollution is the release or presence in the environment of* a harmful / toxic substance. (1)

(b) The lichen grows in unpolluted and polluted air (1); more bricks are covered by the lichen in unpolluted air than in polluted air / the lichen grows better in unpolluted air (1).

2 (a) increase in living standards (1)

(b) two from: land is used for building (1); land is used for quarrying (1); land is used for farming (1); land is used for dumping waste / landfill sites (1)

3 (a) Oxygen levels will decrease (1) because decomposers decompose the dead plants (1) and use up oxygen in the water (faster than it can be replaced) (1).

(b) There will be less oxygen for fish to respire (1); water plants die (1) so less food for the fish / for organisms that the fish feed on (1).

108. Deforestation

1 permanent destruction of forests to make land available for other uses (1)

2 (a) 32 + 34 + 3 = 69% (1)

(b) one from: fuel (1); timber for buildings / furniture (1)

(c) *Planting trees* increases (1) *the rate at which carbon dioxide is taken from the atmosphere. The carbon dioxide is 'locked up' as* wood (1).

3 (a) the variety of all the different species in an ecosystem / area / on Earth (1)

(b) two from: some species are destroyed in the deforestation (1); some species have their habitat destroyed (1); some species have their food sources destroyed (1)

4 three from: peat bogs are a major store of carbon / store more carbon than rainforests (1); peat bogs provide an important habitat for many organisms (1); decomposition of peat releases carbon dioxide into the atmosphere (1); the released carbon dioxide can contribute to global warming (1); there may be a reduction in biodiversity (1)

109. Global warming

1 (a) As carbon dioxide levels increase, the average world temperature increases (1); as carbon dioxide levels decrease, the average world temperature decreases (1); high temperatures occur at about the same times as high carbon dioxide levels (1).

(b) The trend shows a very large increase in carbon dioxide levels (1), which could mean a very large increase in average temperatures (1).

2 (a) methane (1)

(b) *an increase in the Earth's average* temperature (1) *due to rising levels of* greenhouse gases (1)

3 two biological effects of global warming, e.g. change in the migration patterns of birds (1); change in the distribution of species (1); change in biodiversity (1); spread of tropical diseases into new areas (1)

110. Maintaining biodiversity

1 two from: moral / ethical reasons (1); aesthetic reasons, e.g. enjoyment of seeing animals and plants (1); value / usefulness of different species to humans (1)

2 (a) one from: to increase the area of the fields (1); to make it easier for large machinery to move around (1); to reduce weeds / pests that might live in hedgerows (1); farm animals more likely to be kept indoors now (so hedgerows not needed) (1)

(b) biodiversity reduced because (1): two from: fewer plant species (1); fewer animal species (1) food / shelter has been removed (1)

(c) (i) one from: provides shelter for farm animals (1); provides windbreak for crops (1); provides habitat for predators that prey on crop pests (1); may be paid to replant hedgerows (1)

(ii) one from: provides shelter / nesting places for wild animals (1); provides food for wild animals (1); reduces soil erosion (1); provides a safe pathway for animals to move from place to place (1)

3 two from: restores habitat for endangered species (1); reduces the effects of soil erosion because tree roots bind soil together (1); helps to reduce the overall release of carbon dioxide because of photosynthesis by the trees (1)

4 *Hedgehogs may be run over by traffic if they* cross the road looking for food (1); a tunnel allows hedgehogs to move under the road safely (1), so more survive to reproduce (1).

111. Trophic levels

1 (a) herbivore (1); primary consumer (1)

(b) *The grass is a plant, so it is a* producer (1) *which makes* food / glucose by photosynthesis for the deer to eat (1).

2 *clover, 50 kg* (1); mouse, 15 kg (1); snake, 5 kg (1)

3 (a) correctly shaped and labelled pyramid with producers at bottom, then herbivores, then carnivores (1); horizontal widths of bars drawn to scale (e.g. if 1 cm = 250 kg, widths are 10 cm, 2 cm, 0.8 cm) (1)

(b) not all the ingested material is absorbed (1), e.g. it is egested as faeces (1); some absorbed material is lost as waste (1), e.g. carbon dioxide / water in respiration; water / urea in urine (1)

112. Food security

1 (a) having enough food to feed the population (**1**)

(b) *It meets the needs of people today* (**1**), *without reducing* the ability of people to meet their future needs (**1**).

2 two from: cost of fuel (**1**); cost of seed (**1**); cost of fertilisers (**1**); cost of animal feed (**1**)

3 (a) two from: it is more difficult to grow food (**1**); it is more difficult to process food (**1**); it is more difficult to transport food (**1**); supplies of water may be reduced / water may be contaminated (**1**)

(b) The level of food security will be reduced because (**1**) there will be more people to feed (**1**).

4 (a) Table completed, **1 mark** for each row:

Pest management strategy	How it controls pests
introduce a natural predator of the pest	predators attack / kill pests (**1**)
spray insecticides	kills **insect** pests (**1**)

(b) Pests eat / damage crops / animals (**1**); if there are fewer pests, crops yields / quality will be higher (so food security will be improved) (**1**).

5 It would reduce the food supply because (**1**): farmland used for bioethanol instead of food (**1**); plants that could be used to feed people are used to make bioethanol (**1**).

113. Farming techniques

1 (a) Percentage = (100 × 14 900)/(1240 × 1000) (**1**) = 1.2% (**1**)

(b) (i) Energy stored in cow = 1000 – 540 – 150 = 310 kJ (**1**)

(ii) *keeping warm* (**1**) and moving around (**1**)

(c) The efficiency of energy transfer from crops to meat / parts of the cow is small / 31% (**1**); reducing the number of steps in a food chain reduces losses (**1**), so if humans eat crops directly, there is one less step and a more efficient transfer of energy (**1**).

2 The beaker in the box lost less heat / cooled more slowly than the beaker on the bench (**1**), so a chicken indoors would lose heat more slowly than a chicken outside (**1**); a chicken indoors would use less energy to stay warm (**1**), so they would need less food / would transfer a greater proportion to meat / eggs (**1**).

114. Sustainable fisheries

1 (a) It prevents too many fish being caught in a particular area (**1**), so that breeding stocks can be maintained (**1**).

(b) The quota will make little difference to stocks of fish in the North Sea (**1**), because the other boats could continue to catch fish / may catch even more fish (**1**).

2 (a) The proportion of farmed salmon has increased (**1**), from 25% to 70% / by a factor of nearly three times (**1**).

OR

The proportion of wild salmon has decreased (**1**), from 70% to 25% / by a factor of one-third (**1**).

(b) *Farmed salmon are kept in cages, so* have less room to swim / expend less energy (**1**). *As they are provided with food,* they may overeat / not exercise while finding their own food / excess food converted into fat (**1**).

(c) Fewer wild salmon are removed from the sea, so their numbers increase (**1**), but this may lead to less food for other species of fish (**1**), so their numbers decrease (**1**).

3 Young fish are smaller than adult fish (**1**); if the mesh size is too small, young fish are caught / unwanted smaller species are caught (**1**); if the mesh size is large enough, larger maturing / adult fish can escape (**1**) so they remain in the sea and can breed (**1**).

115. Biotechnology and food

1 (a) lean beef (**1**)

(b) Vegetarians do not eat meat and mycoprotein comes from a fungus (**1**); mycoprotein is protein rich / contains just over half the percentage of protein that lean beef does (**1**).

2 (a) glucose / glucose syrup (**1**)

(b) *Air contains* oxygen (**1**). *It is supplied because Fusarium must grow in* aerobic (**1**) *conditions.*

(c) Ammonia contains nitrogen (**1**), which is needed for *Fusarium* to make proteins (**1**).

(d) The fungi release energy because of their respiration (**1**); the fermenter needs to be kept cool so that it does not overheat (**1**), otherwise the enzymes controlling respiration would denature (**1**).

116. Extended response – Ecology

*Answer could include some of the following points to **6 marks**:

Using the graph:

- The rate of deforestation has changed over time
- but does not fall to zero,
- so the deforested area increased over time.

Use of land:

- may be used for farming
- farm animals, e.g. cattle, release methane
- methane is a greenhouse gas.

Use of wood:

- Trees may be burned / used as fuel.
- This releases carbon dioxide.
- Carbon dioxide is a greenhouse gas.

Carbon cycle:

- Fewer trees to absorb carbon dioxide for photosynthesis.
- Dead trees decompose.
- Decomposers release carbon dioxide because of respiration.

Link with climate change:

- Greenhouse gases are a cause of global warming.
- Global warming leads to climate change.

117. Timed Test 1

1 (a) nucleus (**1**)

(b) makes / produces / synthesises proteins (**1**)

(c) two from: bacterial cells have a single loop of chromosomal DNA, plant cells have chromosomes in a nucleus (**1**); bacterial cells have plasmids, plant cells do not (**1**); plant cells have mitochondria, bacterial cells do not (**1**); plant cells have a cell wall made from cellulose, bacterial cell wall is not made of cellulose (**1**)

(d) Length of plant cell = (measured length of cell in mm)/magnification

= 59/500 = 0.118 mm (**1**) = 1.18×10^{-2} m (**1**)

= 1.2×10^{-2} m (to 2 significant figures) (**1**)

2 (a) Petri dishes and culture media will already contain bacteria (**1**); these must be killed so that they do not contaminate the culture (**1**).

(b) heat in a flame (**1**)

(c) secured with sticky tape (**1**) but not all the way round / in a cross (**1**)

(d) Human body temperature is 37 °C (**1**), so incubating at a lower temperature reduces the chance of harmful / pathogenic bacteria growing (**1**).

3 (a) to maintain the milk at a constant temperature (**1**)

(b) one from: to show that milk alone does not cause resazurin to change colour (**1**); to show that it is a process that causes the colour change (**1**)

(c) Live bacteria in pasteurised milk (**1**) grow rapidly at 36 °C (**1**); no live bacteria in sterilised milk (**1**).

4 *Answer should include the following points explained in a clear, logical way, to **6 marks**.

Similarities:

- involve movement of particles
- may be through membranes / into cells / out of cells
- involve concentration gradients.

Differences:

- Diffusion is the net movement from an area of higher concentration to an area of lower concentration / down a concentration gradient, but active transport is movement from a more dilute solution to a more concentrated solution (against a concentration gradient).
- Active transport requires energy from respiration, diffusion does not.

Examples:

- diffusion, e.g. gas exchange of oxygen and carbon dioxide in lungs; absorption of lipids / substances in the small intestine; movement of urea from cells into plasma
- active transport, e.g. movement of mineral ions from soil solution to plant root hairs; absorption of sugars / amino acids in the small intestine.

5 (a) differentiation (**1**)

(b) tissue: group of cells with a similar structure and function (**1**); organ: aggregations / combination of tissues (**1**), each with a specific function (**1**)

(c) Adult stem cells can differentiate / specialise only to form certain type of cells but embryonic stem cells can form many / all types of cell (**1**).

6 (a) It is made in the liver and stored in the gall bladder. (**1**)

(b) Bile neutralises hydrochloric acid from the stomach (**1**) and emulsifies fat to form small droplets / larger surface area of fat (**1**); lipase digests / breaks down fat (**1**) to form glycerol and fatty acids (**1**).

(c) two from: thin (epithelial) surface layer of cells (**1**) gives a short distance for diffusion (**1**); rich blood supply / large numbers of capillaries (**1**) maintains concentration gradient (**1**); large surface area (**1**) means that diffusion happens faster (**1**)

(d) Enzyme activity increases with temperature (**1**) because rate of collisions between substrate and active site increases (**1**); optimum temperature at about 60 °C (**1**); activity decreases above this temperature because enzyme is denatured / active site changes shape (**1**).

7 (a) 8 (**1**)

(b) plasma (**1**); transports carbon dioxide / transports dissolved nutrients / carries cells / platelets round the body (**1**)

(c) transport oxygen (**1**)

(d) platelets (**1**); involved in clotting (**1**)

(e) two from: produce antibodies (**1**) which attach to antigens / pathogens, leading to their destruction (**1**); engulf / ingest pathogens (**1**) and digest / destroy them (**1**); produce antitoxins (**1**) which attach to poisonous substances / make poisonous substances inactive (**1**)

8 (a) any two from: a placebo is a tablet with no active drug (**1**); it is used because sometimes people's health can change if they think only they are being treated (**1**), so the trial needs to look at the difference between people on the drug and people on a placebo (**1**)

(b) The results are likely to be reliable (**1**) because many people are involved in this trial (**1**).

(c) After 4 years the patients taking the placebo were twice as likely to have heart or circulatory disease (6.2 per 100 rather than 3.0–3.4 per 100) (**1**); the study shows that statins help to reduce the likelihood of heart disease (**1**).

9 (a) root hair cells (**1**)

(b) Xylem vessels transport water and dissolved mineral ions, but phloem vessels transport water and dissolved sugars (**1**); xylem vessels transport in one direction (roots → stem → leaves), but phloem vessels transport in either direction (source → sink) (**1**).

(c) removal of anomalous result of 21 mm (**1**); mean value = (59 + 68 + 62)/3 = 63 mm (**1**); 63 mm in 5 minutes, so mean rate of movement = 63/5 = 12.6 mm/min (**1**)

(d) The bubble would move more slowly (**1**); the rate of photosynthesis goes down (**1**), so stomata in the leaves open less widely (**1**).

10 *Answer should include the following points explained in a clear, logical way, to **6 marks** maximum.

Making monoclonal antibodies:

- Mouse lymphocytes / white blood cells stimulated to make a particular antibody.
- Lymphocytes / white blood cells are combined with a tumour cell
- to make a new cell
- called a hybridoma cell.
- Hybridoma cells both divide and make the antibody.
- Single hybridoma cell cloned.
- Large amount of the antibody can be collected and purified.

Uses of monoclonal antibodies (two uses):

- diagnosis such as in pregnancy tests
- to measure the levels of hormones / substances in blood
- to detect pathogens
- to locate or identify specific molecules in a cell or tissue by binding to them with a fluorescent dye
- to treat diseases, e.g. cancer.

11 (a) stops elephants / herbivores eating it (**1**), so photosynthesis / growth is not reduced (**1**)

(b) new flowers are to attract pollinators / bees (**1**); stops ants from competing with pollinators / reducing pollen / nectar / eating the pollen / nectar (**1**)

12 (a) carbon dioxide + water → glucose + oxygen (**1**)

(b) energy / light is transferred (**1**) from the surroundings (**1**)

(c) two from: used for respiration (**1**); converted into insoluble starch for storage (**1**); used to produce fat / oil for storage (**1**); used to produce cellulose (for the cell wall) (**1**); used to produce amino acids for protein synthesis (**1**)

(d) Rate of photosynthesis increases as the light intensity is increased (**1**) but reaches a maximum rate (**1**) when other factors / carbon dioxide / temperature / amount of chlorophyll become limiting factors (**1**).

(e) *Answer should include the following points explained in a clear, logical way, to **6 marks** maximum.

- Cells contain chloroplasts / chlorophyll.
- Chloroplasts are the site of photosynthesis.
- Cells in upper part / palisade mesophyll have cylindrical shape
- so pack closely together / few cell walls for light to pass through
- have more chloroplasts.
- Cells in lower part / spongy mesophyll are irregularly packed
- which provides air spaces / large surface area
- for gas exchange.
- Stomata in lower surface / epidermis
- allows gas exchange.
- Size of opening controlled by guard cells
- to control loss of water vapour / controls rate of transpiration.

13 (a) glucose → ethanol + carbon dioxide (**1**)

(b) one from: making bread (**1**); making alcoholic drinks (**1**); making bioethanol (**1**)

14 (a) Muscle cells need more oxygen (and produce more carbon dioxide) (**1**), so breathing rate increases to obtain more oxygen / increase rate of gas exchange in the lungs (**1**).

(b) Respiration is an exothermic process / transfers energy to the surroundings by heating. (**1**)

(c) During long exercise anaerobic respiration takes place in muscles (**1**); lactic acid is produced in the muscle cells (**1**); after exercise, extra oxygen is needed to react with the lactic acid (so it is removed) (**1**).

122. Timed Test 2

1 Sensory neurones transfer the impulse to the relay neurone / spinal cord (**1**); relay neurone / spinal cord / CNS transfers the impulse to a motor neurone (**1**); motor neurones transfer the impulse to the effector / muscle (**1**).

2 (a) (350 + 330 + 340 + 320)/4 (**1**) = 335 ms (**1**)

(b) 260 ms reading is anomalous / out of line with others (**1**), so it was excluded when the mean was calculated (**1**).

(c) (The conclusion is not good because) three from: mean reaction time for boys was less than the mean reaction time for the girls (**1**); but the difference is very small (**1**); it could be random / due to chance (**1**); the sample size was very small (**1**); the result may not be repeated across a larger sample (**1**).

3 (a) three from: contains light-sensitive receptor cells (**1**); rods sensitive to low light levels (**1**); cones sensitive to bright light / colours (**1**); connected to the optic nerve (**1**)

(b) Ciliary muscles relax (**1**); suspensory ligaments are pulled tight (**1**); lens is pulled thin (so it only slightly refracts light rays) (**1**).

4 (a) glucagon released (**1**); stored glycogen converted into glucose (**1**); blood glucose levels increase (**1**)

(b) to reduce glucose concentration in the blood to minimum levels (**1**)

(c) (The person does not have diabetes because) three from: starting level is within normal limits (**1**); in the test, the blood glucose concentration rises only to 8 / does not rise above 9 (**1**); it then falls to normal levels (**1**); this can be brought about only by insulin release (**1**).

5 (a) liver (**1**)

(b) amino acids deaminated to form ammonia (**1**); ammonia converted to urea (**1**)

(c) permeability of kidney tubules decreased (**1**), so less water reabsorbed in the kidneys / nephron / collecting duct (**1**); less water in the blood (**1**)

6 (a) FSH: causes maturation of an egg in the ovary (**1**); LH: stimulates the release of the egg (**1**)

(b) inhibits the release of FSH (**1**); stimulates the release of LH (**1**)

(c) If thyroxine levels fall, more TSH is released (**1**) by the pituitary (**1**), so thyroxine levels rise / more thyroxine released by the thyroid (**1**); reverse situation happens if thyroxine levels rise (**1**).

7 (a) four from: positive gravitropism by shoot (**1**); negative gravitropism by root (**1**); auxins move to lower part of the shoot / root (**1**); cells in lower side of shoot elongate more (**1**); cells in upper side of root elongate more (**1**)

(b) two from: ending seed dormancy (**1**); promoting / starting flowering (**1**); increasing size of fruit (**1**)

8 (a) 23 (**1**)

(b) Copies of the genetic information are made (**1**); cell divides twice to form four gametes (**1**); each with a single set of chromosomes / gametes are genetically different from each other (**1**).

(c) Completed Punnett square diagram, e.g.

		Mother	
		X	X
Father	X	XX	XX
	Y	XY	XY

mother XX (**1**); father XY (**1**); children completed correctly (**1**); probability = 50% / 1:1 / 1 in 2 (**1**)

9 Completed Punnett square diagram, e.g.

		Mother	
		G	g
Father	G	GG	Gg
	g	Gg	gg

parents' alleles (**1**); children completed correctly (**1**); gg has disease (**1**); GG is not a carrier (**1**); probability = 25% / 1:3 / 1 in 4 (**1**)

10 (a) four from: a polymer (**1**); made up of two strands (**1**); forming a double helix (**1**); four different nucleotides (**1**); nucleotides consist of common sugar and phosphate group with a base attached to the sugar (**1**); bases are A, C, G, T (**1**)

(b) TAGCCA (**1**)

(c) $330 \times 3 = 990$ (**1**)

(d) Template (leaves the nucleus and) binds to a ribosome (**1**); carrier molecule brings a specific amino acid (**1**); amino acid added to growing protein chain (**1**); protein chain folds up when it is complete (**1**).

11 (a) Giraffe population has a variety of neck lengths (**1**); those with the longest necks can reach the most food (**1**), and so survive to reproduce more often than the shorter-necked ones (**1**); alleles for longer necks passed to offspring in the next generation (**1**).

(b) two from: the theory challenged the idea that God made all the animals and plants (**1**); there was not enough evidence at the time to convince many scientists (**1**); the mechanism of inheritance and variation was not known until many years after the theory was published (**1**)

12 (a) not extinct because: extinction is when there are no remaining individuals of a species still alive (**1**), but lemurs still exist on Madagascar (**1**)

(b) two from: monkeys passed on diseases to lemurs (**1**); monkeys killed the lemurs (**1**); monkeys were more successful in competing for resources / habitat (**1**)

(c) one from: fossil record is incomplete (**1**); not all parts of organisms fossilise (**1**); geological activity destroys fossils (**1**)

13 three from: large feet to stop sinking in sand (**1**); nostrils that close to stop sand getting in (**1**); fat store in hump for energy (**1**); fat stored in hump rather than the rest of the body to reduce insulation (**1**); long legs to keep the body away from hot sand (**1**)

14 (a) bird (**1**)

(b) sunlight (**1**)

(c) energy lost through faeces / respiration / movement / keeping warm / other life processes (**1**); not all the organism can be consumed / digested / absorbed by the next organism (**1**)

(d) four bars getting narrower from bottom to top (**1**); all bars labelled (**1**)

(e) Hedgehogs and birds compete to eat slugs (**1**), meaning there will be fewer birds (**1**), so fox numbers decrease (**1**).

15 five from: quadrats to sample areas (**1**); transect at right angles to the road (**1**); repeat with several transects at different places along the road (**1**); calculate mean numbers of plants at each distance (**1**); anomalous results identified and accounted for (**1**); same sized quadrats (**1**)

16 *Answer should include the following points explained in a clear, logical way, to **6 marks** maximum.

Disadvantages of salmon farming:

- Uneaten food can cause pollution.
- Salmon produce waste and this can cause pollution.
- Pollution may lead to local species dying out.
- Diseases (e.g. sea lice) from the farmed salmon can transfer to wild fish, and harm or even kill them.

Advantages of salmon farming:

- provides employment
- reduces fishing of wild fish.

Advantages of mussel farming:

- No food is added to the water
- so no uneaten food to cause pollution.
- Mussels remove waste from the water.

Disadvantages of both:

- reduces biodiversity
- can spoil the landscape.

Notes

Notes

Notes

Notes

Published by Pearson Education Limited, 80 Strand, London, WC2R 0RL.

www.pearsonschoolsandfecolleges.co.uk

Typeset, illustrated and produced by Phoenix Photosetting
Cover illustration by Miriam Sturdee

First published 2017

20 19 18 17
10 9 8 7 6 5 4 3 2 1

British Library Cataloguing in Publication Data
A catalogue record for this book is available from the British Library

ISBN 978 1 292 13501 4

Printed in Slovakia by Neografia

Acknowledgements
The publishers are grateful to Sue Kearsey, Stephen Hoare and Gary Skinner for their help and advice with this book.

The author and publisher would like to thank the following individuals and organisations for permission to reproduce their photographs:

Science Photo Library Ltd: BioPhoto Associates 1, Steve Gschmeissner 6, 9.

All other images © Pearson Education

Note from the publisher
Pearson has robust editorial processes, including answer and fact checks, to ensure the accuracy of the content in this publication, and every effort is made to ensure this publication is free of errors. We are, however, only human, and occasionally errors do occur. Pearson is not liable for any misunderstandings that arise as a result of errors in this publication, but it is our priority to ensure that the content is accurate. If you spot an error, please do contact us at resourcescorrections@pearson.com so we can make sure it is corrected.